T0089000

# Process Plant Lifecycle Information Management

ROBERT YANG

iUniverse, Inc.
New York  Bloomington

Process Plant Lifecycle Information Management



iUniverse books may be ordered through booksellers or by contacting:

iUniverse
1663 Liberty Drive
Bloomington, IN 47403
www.iuniverse.com
1-800-Authors (1-800-288-4677)



ISBN: 978-1-4401-4757-9 (pbk)
ISBN: 978-1-4401-4758-6 (ebk)

Printed in the United States of America

iUniverse rev. date: 8/3/2009

# CONTENTS

# Location of Figures and Tables

# INTRODUCTION

Designing, building, and operating process plants all involve complicated tasks. In a plant lifecycle, the engineer or designer translates the owner's the set of processes that meets the owner's production requirements into a set of technical information. Procurement and construction then converts such information into assets that represent the physical process plant. From there, the as-built information provides the operating procedure for the process plant, which operates and is maintained by owners and operators to produce the required products. Stated differently, the key elements of plant lifecycle ultimately devolve to generation, manipulation, and management of information.

In the traditional way, documents are still used to represent information. With the rapid advancement of electronic documents, there has been a major jump of productivity in terms of generation, distribution, usage, and collaboration of information. In fact, the trend now is, however, to emphasis more on data-centric rather than document-centric representation of information. The usage of database approach is another level of productivity increase because information can be indexed, integrated, stored, retrieved, and managed more efficiently. Theoretically, documents can all be generated from their databases.

Three-dimensional model represents a third important information source. Significantly, in digital representations there not only is no scale factor because it is the digital representation of the actual plant but it also contains all dimensional information, which can be linked to documents and databases for all components represented by tags in the plant. Throughout a plant lifecycle, 3-D models are used for engineering and design, procurement, construction, and operations, and maintenance of the plant.

Efficient plant lifecycle information management has to satisfy three basic requirements: what, when and how information to be managed. Considering the large amount of information in a process plant, one must understand, plan, and manage that for each step of the lifecycle, to determine not only what information needs to be shared, integrated, and used throughout lifecycle of the plant but even more important to understand *when* such information is required and to be managed. Because changes occur continuously throughout the plant lifecycle, timing of available right information is critical to avoid confusion and potential misuse of the wrong information.

To manage information successfully, it is important to have a single source of information throughout the plant lifecycle. The providers of information and the users of information are all connected and can share the same information through that single source. This eliminates the complication of multiple information sources and the potential mistake of using redundant information. One type of single-source approach is an *information data warehouse* in which database information is organized, indexed, and integrated through linkage to documents and 3-D models of the process plant.

Information integrity is another important consideration for information management. Information integrity means information accuracy and currency. The users of information must have confidence that the stored, single-source information is correct and appropriately up to date. Modern technology can store very large quantity of digital information and it can be indexed, content-managed, and retrieved efficiently.

Modern process plants also have to satisfy more safety and environmental requirements. Lifecycle information can be collected and assembled to meet compliances. The plant is designed to comply with the design specification, and the plant then is built as designed. Information is tracked, stored, and retrieved for operational readiness reviews and license applications for operation of the plant.

This book is not a 'how to' book on information management. Considering the complexity and different types of process plants, it is neither practical nor possible to take such an approach. This book shows the basic principles and methods of process plant lifecycle information management. While there are many different terms and definitions used in the book, the readers should be able to recognize and understand their meanings from their knowledge and experience.

Other than what, when, and how to manage plant lifecycle information, we have to ask the basic question of why we want to do it. At the project level, the obvious answer is to increase productivity so plant capital costs and time to market can be reduced. At the plant level, the answer is to reduce operational expense and to maximize time in market. With proper information and information management, the most important reason is that the owner and operator now have a tool to optimize operating parameters to improve both the quality and the quantity of the process plant products.

We conclude this book with a chapter on implementation of plant lifecycle information management. There have been many technology tools developed recently for information management. Consequently, it becomes necessary to decide which one best fits a particular situation or if one is too complicated to justify for the current project. In addition to technology, the other two elements to successful implementation are people and work processes. Work process is the most difficult to handle because it usually requires an iteration process first to find the right choice. Technology cannot simply be forced to fit into a traditional work process; rather, the work process must change to take advantage of the technology.

We have included two appendices in the book. One is on several topics related to information management. The other is on the subject of building or facility lifecycle information management. While there are more buildings and facilities, the information management application is at least few years behind compared to process plants. Building Information Modeling (BIM) is the recent push in that direction. And though application concepts and approaches are similar, there are major differences and the subject belongs to a different book.

Plant lifecycle information management—and more specifically the use of a data warehouse—has been implemented and used by EPCs (engineering, procurement, and construction companies) and owners and operators of different process industrials at various stages of plant lifecycle applications. Technology companies, in addition, have developed various information management tools that are not 'pie in the sky' but have been shown to have specific successful applications. Because development and implementation of technology require interactions of knowledge and experience from many sources, we are unable to acknowledge any specific individuals and companies on their contributions to the field of plant lifecycle information management but to say 'thanks' to all.

# CHAPTER I
# PROCESS PLANT

## 1.1 Process Plant Types

As the process plant name implies, it is a plant that involved in one or more processes. It can be a chemical plant that combines chemical components or compounds into chemical products. It can be an oil refinery that refines crude oil into different types of petroleum products. A gas plant cleans and adds components to the natural gas from the ground before it feeds to a piping network as fuels.

In the food industry, agricultural products, plants, and livestock are converted into food products like packaged frozen food, dairy products, candies and so forth. A winery or a brewery makes alcoholic drinks through the fermentation process. Other process plants make various soft drinks and bottle drinking water.

The pharmaceutical industry is another major process industry. Biotechnology companies operate process plants that deal with human cells, viruses, and so forth to develop new medicines to cure diseases.

The federal government operates many different types of process plants. The Department of Energy processes nuclear fuel, cleans up nuclear waste, and operates alternative energy plants. The government also operates process plants to manufacture explosives, ammunition, and various weapons. In recent years, major process plants to decommission chemical weapons have been built.

On the state and municipal level, there are thousands of wastewater and water treatment plants. These are process plants that treat wastewater and provide clean water for the safe and convenient life of all citizens.

Electrical utility companies operate power plants that generate electrical power from coal, oil, gas, or nuclear resources to support industries, homes, and cities. Basically, all of these are process plants that convert fuel energy to heat energy and through a process media and mechanical equipment generate electrical power.

In the mining industry, production of copper and silver involves production of the basic metal through a reduction process. The plant may use extensive material handling equipment, but the basic production function is a process plant.

As may be seen, there are many types of process plants covering a wide spectrum of industries. The term 'process' should be understood in a very wide sense. This covers a huge business area. The potential capital investment for new plants and upgrade of existing plants is in the trillions of dollars. The yearly costs of operations and maintenance of such plants are in the same magnitude of dollars. If we can reduce these costs through information management by a very small amount, even only 0.1% of the total investment, the potential savings comes to billions of dollars.

## 1.2 Process Plant Features

In a process plant, it is the owner's or operator's responsibility to select a process, whether it is proprietary or licensed. The process defines many parameters: process reactions, streams, compositions, temperature, pressure, flow rate, and so forth. In engineering terms, these parameters are converted from process and process diagrams into Piping and Instrument Diagrams (P&IDs). In other words, every process plant has P&IDs.

Process plants require extensive instrumentation to measure process performance. From those measurements, control systems are designed to allow delicate control of the process. In many plants, more measurements and controls are required to ensure safe operations of the plant. In the last twenty and thirty years, P&IDs have steadily grown in complexity to accommodate increased instrumentation and control system requirements. Digital systems and computers have added even more complexity and ability to handle such complexity.

Piping is the usual mechanical system that contains and directs process flow media from one process operation to the other. Piping contains liquid or gas or a combination of these—at pressures from vacuum to a few thousand pounds per square inch and temperatures up to a few thousand degrees. Piping of various sizes and lengths made from common carbon steel to exotic alloy steel is used to contain various flow media and chemicals. For some process plants, stainless steel tubing and piping with special coatings are required to meet cleanness requirement.

Process equipment consists of reactors, separators, heaters, exchangers, vessels, pumps, and so forth. For some processes, equipment must be specially designed and fabricated. Equipment may be relatively small or very large, such as vessels 15'–20' in diameter or columns 50'–60' high and walls 10"–12" thick. For high temperature and corrosive process mediums special alloy materials may be required.

In some process plants, large sites are required to accommodate storage tanks, concrete pools, and ponds. In such cases, instead of pipes, concrete channels may be used to content the flow media.

## 1.3 Process Plant Support Infrastructure

To facilitate process flow and maintenance, piping discipline usually lays out major process equipment in an equipment plot plan. This determines the general physical size of the process plant site. Depending on the type of process plant, process equipment may be located outside as in a refinery or in inside as in a pharmaceutical plant. The specific needs of the process will dictate this.

A civil engineering team develops the plant site, the roads and support utilities—such as electricity, water, and sewage connection—from the outside boundary to the plant. A structural engineering team designs steel supports for pipes, concrete foundations for process equipment, and any outside concrete structure, such as a settling pond.

An architectural team does the building layout such as control rooms and the process equipment arrangement inside the building, while the structural team sizes the building components. Inside the building, a mechanical engineering team designs the plumbing, HVAC, fire protection, and any lifting equipment requirements. They may also design any special requirements—such as HEPA air filter, cascade airflow, vibration isolation, and so forth—if required.

An electrical engineering team designs the power access for process equipment, controls, and instrumentation and sizes electrical equipment—such as motor control center, switch gears, and so forth—and other requirements to support utilities. In some cases, utility instrumentation and control and communication equipment are also a part of electrical requirements.

## 1.4 Lifecycle Steps

Traditionally, the lifecycle of a process plant is divided into six steps. These are:

1. Process development

2. Front-end engineering
3. Detailed engineering
4. Procurement
5. Construction
6. Operations and maintenance

See Figure 1.1.

## Figure 1.1 Plant Life Cycle -Steps

INFORMATION INTEGRATION

INFORMATION MANAGEMENT

**1** - Process **2** - Front End Engineering
**3** - Detailed Engineering **4** - Procurement
**5** - Construction **6** - Operations/Maintenance

Usually, in steps 1 and 2 the owners or operators dictate the process requirements and perform some of the front engineering studies to show feasibility and cost justification of developing such a plant. This may include process product specifications, projected performance requirements, estimated plant cost and schedule, projected operational cost, and other elements, as required

Step 3 to 5 are usually included in an EPC (engineering, procurement, construction) phase and are handled by an EPC contractor or by the owner directly. This may also be called the capital project (CAPEX) phase of the plant. After construction, mechanical completion, and start-up testing of the plant, the owners or operators begin to operate and produce the product from the plant. Step 6 is usually referred to as the operational phase (OPEX) of the plant lifecycle.

Figure 1.2 considers the plant lifecycle from a time schedule and cost perspective. For a typical process plant lifecycle, the duration of EPC phase may be two to three years while the operational phase may be twenty years or longer. Since the cost of the EPC phase may represent 20–25 percent of the total plant lifecycle cost, this shows the importance of managing operational costs because a potential saving of 10–12 percent per year in operation can essentially pay for the EPC phase of the plant. As we'll see, effective information management in operations and maintenance can achieve this result.

**Figure 1.2 Plant Life Cycle – Schedule Time & Costs**

INFORMATION INTEGRATION

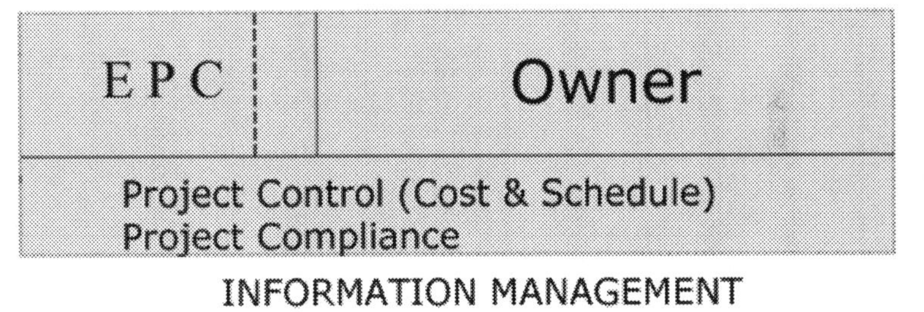


EPC: 2 to 3 years, 15% to 30% lifecycle costs
Owner: 15 to 20 years, 65% to 85% lifecycle costs

The above descriptions of plant lifecycle in discrete steps are misleading in a sense. In the early days of process plant development, some owners—including government agencies—considered engineering/design and construction as two separate parts and would not allow one contractor to handle both functions because of potential conflicts of interest. Today, this is no longer true. The potential cost savings from integration and information management alone would allow one EPC contractor to handle complete plant development from engineering through construction. Even today, however, there is still a sharp discontinuity between EPC phase and operational phase. The so-called hand over is still a major issue. The owner typically separates CAPEX and OPEX phase on different budgets and through different organizations. There is also the issue of how to integrate and handle the responsibility and commercial terms from the EPC to the owner's or operator's company.

From an information management point of view, the plant lifecycle should be a continuous process. The information flows efficiently from one operation to another with inputs and feedbacks from each subsequent operational step. Within engineering, all disciplines work is ideally performed in an integrated and corporative manner. The owner provides information on the operational and maintenance requirements as early as possible of the project. The procurement team coordinates with engineering and performs purchasing functions on the most cost effective way to define materials and equipment requirements. The construction team participates early in the project to look at the constructability and then jointly develops a construction plan with engineering.

On the other hand, there is a logical difference between the EPC phase and the operational phase. While the EPC phase generates plant information that defines the performance and physical configuration of the plant, the operational phase uses such information for production, efficient operation, and maintenance of the plant. For brevity, the EPC phase can be called the project phase while actual plant operations may be called the operational phase.

## 1.5 Project Phase Functions

As shown in Figure 1.3, project execution is the main EPC function starting with the desired process itself, and then moving through preliminary and detailed engineering, procurement, construction, to end at mechanical completion through systemization and testing. At the completion of the EPC phase, the project is ready to be turned over to the owner or operator for start-up of operations.

**Figure 1.3 Project Functions**

| Project Control | Project Execution | Project Compliance |
|---|---|---|
| Schedule | Process | Configuration (Change) Management |
| Cost | Preliminary Eng/Design | QA/QC (verification) |
| WBS | Detail Eng/Design | Safety |
| | Procurement | Environment |
| | Construction | |
| | Mechanical Completion | |
| | Start-Up | |

The engineering step of the project generates all of the plant technical information that describes and defines the performance and configuration of the plant. Procurement buys equipment, components, and materials as specified by engineering and prepares subcontracts for fabrication and other functions if required. Procurement also generates information via acquired vendor data on the physical components of the plant. Construction uses the procured physical components and commodity items to build the physical plant that satisfies both the specified engineering design requirements and the operational requirements of the plant through checkout and testing.

Project execution is supported by project control and project compliance functions. There is continuous flow of information across these functions. Project control extracts information from EPC functions and plans and manages cost and schedule functions on the execution of the project. There are many available cost and schedule control programs that satisfy this part of project function. Frequently, a work breakdown structure (WBS) identifies and tracks cost and schedules the various steps of the EPC project functions.

Project compliance is one of the critical requirements of project functions that it also extracts information from project executions to verify their compliances. It shows that the plant is designed to satisfy the owner's plant performance specifications. Equipment, control systems, materials, and so forth are purchased to meet the specified design specifications, and the plant is then constructed as designed. Safety, environmental, and government regulations are also satisfied, and there must be sufficient documentation and information to verify such compliances. There are many stories of very costly plant start-up delays because a plant cannot obtain licenses to start operation for having failed to comply with the plant operational requirements.

## 1.6 Plant Operational Phase Functions

After mechanical completion and start-up testing, the plant is ready for production and operations. The main operational and maintenance functions of plant operation are shown in Figure 1.4.

**Figure 1.4 Plant Functions**

| Plant Support | Plant Functions | Plant Compliance |
|---|---|---|
| Engineering | Operations | Configuration (Change) |
| Procurement | (P&ID,DCS,3-D) | Management |
| Cost | Maintenance | QA/QC (verification) |
| Schedule | (Program) | Safety |
| IT | Asset Management | Environmental |
| | Information Management | |
| | (Server) | |
| | Document Management | |

Process performance related information such as P&IDs, digital control system settings, safety control systems, and so forth are the first group of information required for operations. (Process-related equipment and instrument datasheets are available for easy access for operations.) Information on electrical, mechanical, and safety support equipment and systems are the next level required for operations. Information on physical configuration of support infrastructure is not as critical for daily operation but is required for maintenance and potential revision of the plant.

There are many available maintenance programs. But the starting point must always be a good asset management program. Asset management concerns not only management of physical assets but also management of asset information on process equipment, mechanical equipment, electrical equipment, instruments, and control components. A good document management system and an information management system—such as an information data warehouse—relates to maintenance as well if one is set-up for the plant.

Other plant support functions include field engineering, field procurement, programs to manage plant cost and schedule, and information technology (IT) functions. Plant compliance also continues to be important to meet satisfactorily plant safety and environmental regulations in the event of plant changes.

# CHAPTER 2
# INFORMATION MANAGEMENT APPROACHES

## 2.1 Overview

Process plants have been designed, built, and operated successfully for many years. In the old days, the owner's considerations when selecting an EPC to design and build a plant were based on:

- How many similar plants have they designed and built
- Do they have qualified technical people to put on the project who have worked together before as a team?

In other words, to successfully design and build a plant depends on the knowledge and experience of the EPC contractor and if that experience can be transferred to the current project. Communication between the owner and the EPC relies on understanding and trust.

If we take a backward look at the lifecycle information requirements of a plant from the owner's or operator's point of view, these involve:

- How the plant can be operated successfully and efficiently to produce the right product once it is built, and how the plant can be maintained efficiently?
- Proven project management skills
- What criteria should the EPC contractor be provided for the project?
- What are the most cost-effective ways for the EPC to generate,

organize, distribute, control, and manage the information for the plant design and construction?

- Once the plant is built, how will the EPC effectively hand over the as-built information to the owner?

One critical issue of information management is what and when information needs to be managed. There are many potential approaches that depend on work processes, types of information, information usage, and when information is available. One developed approach is called 'two levels of information management' described in a subsequent section.

Information integrity is the key requirement to successful information management. Information integrity means having a single source of information with guaranteed accuracy and currency. Single-source information avoids confusion and potential mistakes from generating or dealing with redundant information. The users can have confidence that the single-source information is accurate as per specified date.

Information management requires efficient storage, retrieval capability, distribution, and viewing of information. As there are many changes throughout the plant lifecycle, information change management is another critical requirement. Flowing of information from each step of plant lifecycle requires transferring of information from one group to the other group, which means integration of information. Integration involves linking various types of information as defined by the inputs and outputs requirements of different groups. In addition, coordination and collaboration are required for sharing of information throughout the steps of plant lifecycle. We know that automation increases productivity, but we cannot have automation without first integrating information. Automation requires the development of rules, and sometimes the rules can become so complex that they cancel the advantages of automation.

As we can see, information management addresses three basic issues: what information needs to be managed, when information needs to be managed, and how information is managed. We examine these in more detail next.

## 2.2 Information Types

**Document - Paper**

Paper documents are historically the means to record, distribute, transfer, and store technical information. In engineering school, the first thing a student learns is how to read engineering drawings. Then, as she or he enters the design engineering profession, the requirements of other documents—like standards, calculations, criteria, specifications, datasheets, and so forth—to support the engineering drawings are learned.

A main shortcoming of paper documentation is its information is subject to interpretation by the user of the document. Rules and guidelines may have been established to help users to read and interpret documents correctly but omissions and mistakes inevitably occur. Also, once paper documents are produced, they are cumbersome to change, reproduce, and distribute. Accessing the specific desired information may also be unnecessarily difficult if the document is very large or complicated. Paper documents may also contain duplicated or redundant information, and there is no efficient means to verify that information in or between several documents is consistent. Changes in one version may not be reflected in others.

The classic model for paper document organization is the library, but this highly organized form of retrieval, distribution, and storage is largely static and poorly suited to the dynamic, highly interrelated situation of the process plant. This is primarily because paper documents have no 'intelligence' that allows the content of the document to be organized and managed.

**Document - Electronic**

With the rapid advancement of computers in the last three decades, electronic documents generated by computers have essentially resolved most of the problems of paper documents. Computer aided design/drafting (CAD) has changed completely the generation of engineering drawings. It not only produces drawings efficiently but changes the basic approach to engineering design. Word processing and spreadsheet programs similarly have changed the production of other types of technical documents. These changes imply that electronic documentation would also change the basic approach of technical information management, and this is indeed the case.

As electronic files, electronic documents can be identified, stored, and managed by computers through a document management system. Distribution of electronic documents through network is fast and accurate. Electronic documents can be made 'intelligent' so that the content information can be extracted and organized for specific usage (such as checking for accuracy or redundancy). Electronic documents can be linked electronically with other documents for information flow, integration, and verification. Electronic documents managed by a good document management system provide a stepwise increase of efficiency in information management.

**Data**

Data accessed through databases and database programs is the most efficient type of information. Information can be obtained directly from data for specific application. It radically improves every step of information management from retrieval, publication, distribution, and storage. Theoretically, any documents could be generated from data in databases. Piping ISOs, for instance, could be generated from programs driven by their databases. Documents such as valve listings, instrument datasheets, equipment specifications, and so forth could similarly be produced from developed templates with provisions to fill in the data from their databases.

Because information can be linked readily from databases, this leads to a data warehouse approach where all database information is linked and stored in a single, centralized location. Data is the basic requirement for automation through rule-based approaches. There are, however, several considerations before one can take full advantages of data applications. We need to specify more exactly the data requirements for each step of work requirements because there essentially is no (and will not be any) interpretation of information required. The legacy approach to work processes also must change to properly implement a data-centric approach to information and information management.

At this point, we would like to mention the more extensive use of 'tags' as a data approach in the engineering/design of a process plant. A 'tag' is defined as an 'intelligent' attributed object in an attached database that specifies its properties. A tag is an engineering item that describes the engineering requirements at a given plant location. It facilities linking of information from database to document to 3-D model.

**Combined Documents and Data Information System**

Even though we prefer a data-centric approach to information and information management, the practical approach of today is still heavily based on an information system of documents, particularly in long-established process plants. Electronic documents are increasingly replacing paper documents especially in the technical field. But, in many cases, paper documents are still required as legal documents for drawings, equipment specifications, operating procedures, and so forth.

Electronic documents are generated through CAD, word processing, spreadsheets, and other application programs. Some paper-based documents, such as vendor data, construction documents, and so forth can be scanned as electronic files. Documents then in the electronic format are stored, retrieved, distributed, viewed, and used efficiently through a document management system.

Data generated from design programs are assembled as databases in electronic files. Information is also extracted from intelligent documents and organized as databases in electronic files. All such electronic files can be organized and managed through a document management system.

As one can see, a combined documents and database system is most likely the information type to be used for information and information management of today's process plant lifecycles.

**3-D Models**

Another information type that has tremendous impact on the success of plant information management is the 3-D model. In the last ten to fifteen years, plant 3-D models have been developed and have come to be accepted as powerful tools in engineering and design, construction, and even operations and maintenance for process plants. An important concept is that a computer generated plant 3-D model is not a scale model of the plant but is an electronic and digital representation of the actual plant. Depending on the complexity of the model, it contains all of the configuration and dimensional size information of the plant, including mechanical, structural, or electrical designs for components within the plant as well. Furthermore, all attribute data information —such as structural beam size, electrical conduit length, and so forth—used to construct the model can be retrieved from the model. Additional databases—represented as tags—may be attached to plant components and systems.

In engineering and design, the 3-D model is the focal point for configuration management and multi-discipline coordination. Automatic interference detection is a powerful tool to minimize changes both in the design and in the field during construction. A 3-D model also provides information for construction planning, construction sequencing, and progress reporting. There are also many uses of 3-D models in operations and maintenance. They are used for operation sequencing and monitoring, safety system setting, operator training, trouble shooting, emergency responses, and setting of maintenance priority.

If properly constructed and linked to the databases, 3-D models contain all of the essential information that represents the actual plant.

## 2.3 Information Management Requirement

Throughout the lifecycle of a process plant, information is generated, distributed, and used by each subsequent group in the lifecycle. The first question is what information is required from each group. Depending on the type of plant and the work processes involved, information requirements are not always the same. However, for better information management, the requirements need to be well defined and coordinated from each lifecycle group as early as possible. Clear understandings of information inputs and outputs from each work step should be defined so proper information can be organized for sharing by all work groups. This is part of information coordination and collaboration. The flow of information from the output of one group to the input of other group is linked, and so this is a part of information integration.

The critical element of information management is the assurance of information integrity. Information integrity means accuracy and currency. As information is generated and published from one group to be used by the subsequent groups, the provider of information has to guarantee that the published information is accurate. But, the published information may change with time, so a time element also must be associated with published information and be documented as different versions or updates with publication dates. In other words, the users of information need to have the ability to access published, accurate, current information.

One way to assure information integrity is a single source of controlled and managed information. Considering different information types, a centralized data warehouse that is linked to various application programs, document management, and data management program is needed over the entire course of the plant lifecycle.

Another critical problem of information management is the requirement to have standard and common information formats. Engineering and design programs generate information in different formats. Vendors also provide information in their preferred formats even that may be specified in the procurement process. To satisfy their operations and maintenance program, owners or operators may have their own information formats as well. There are many industrial specifications on information standards and formats but they may not always fit a given situation. Translation of information from one format to the other is cumbersome and costly. The preferred approach is that participants of the project from the very beginning should coordinate and agree as early as possible on engineering and design application programs, the information management program, and the information standards and formats.

As we'll see with respect to project compliance, proper information storage is yet another critical requirement. Because storing large quantities of electronic information is no longer an expensive undertaking, all changes of published and controlled information throughout the plant lifecycle can be stored, tracked, and retrieved for use if necessary. In the beginning of the project, proper planning of information storage and efficient retrieval of such information can pay off very well in the overall information management.

## 2.4 Two Levels of Information Management

With the rapid development of high-powered computers and database programs, there is an increasing rate of information generation and a more rapid rate of change of that information. These make effective management of information increasingly difficult.

For example, at the height of project design development, downstream design discipline such as electrical requires input information from the P&IDs and process equipment discipline. As changes occur from one design session to the next, the electrical team must adjust the previous session's information to conform to the newly changed information from the upstream disciplines. A related but more serious problem is the difficulty or inability in following and replicating such changing information continuously.

Another issue concerns how design calculations from one discipline are more conveniently left with the discipline that generated them. For example, information generated by the process team is almost never of any direct interest to the structural team, and vice versa. Nevertheless, there will be critical information that will be needed, and will be needed to be shared, between the two disciplines. As such, the sharing of information, as well as distinguishing between essential and inessential information, must be managed as well. This will apply to all elements of a process plant throughout its lifecycle.

We can see, therefore, that there are two levels of information. One level, which we call information as information in a working environment (WE), is information being continuously developed in a work process and is usually confined to the discipline that generates it. On the other hand, information to be shared or used by many disciplines is usually called 'posted' or 'recorded' information. We call this level of information as information in in a controlled environment (ICE).

WE is information generated and revised continuously on a daily or even hourly basis within a working discipline. If we draw a boundary around the discipline, WE information is confined within that discipline. For example, the piping discipline develops a plot plan from the arrangement of process equipment and other components. Subsequently updated plot plan information is later released either as controlled information for use within the discipline or as posted information in ICE for use by the structural, instrument, and electrical discipline teams.

Information in ICE is guaranteed to be accurate by the provider of that information. Information in ICE is up-to-date as of the specified current date by the provider of that information. In other words, information in ICE is guaranteed to have information integrity. Information in ICE is organized for use and sharing by the other users according to their needs. It can be viewed, published, and extracted efficiently as inputs to generate other documents if so required. Information in ICE can only be changed by the provider of that information.

For example, the mechanical process discipline (P&IDs) has information relationship with inputs from the process discipline, and outputs to the piping, instrument, and process equipment discipline. In the early design development of P&IDs, the mechanical process discipline posts certain P&IDs on every Friday afternoon as ICE information. On Monday morning, all related disciplines use the P&IDs in the ICE as the current input information for their design that week. In the meantime, P&IDs are being changed continuously during the week by the mechanical process discipline in light of new process input information in their WE, but this has no impact on the posted P&IDs in the ICE. New P&IDs with all changes identified are again posted as ICE on next Friday to be used by the downstream disciplines. This process continues until the design development reaches a point where there are few changes. Then, the posting may be changed from weekly on every Friday to a cycle of every two to three weeks. This illustrates a formal management of information from WE to ICE, but it does not prevent informal exchanges of information between disciplines during the design development stage.

Vendor searching, qualification, and evaluation are usually considered in the WE functions of the procurement discipline. By contrast, purchase orders issued by procurement are posted as ICE because they entail relationships with engineering, construction, and ultimately operations and maintenance by the owners or operators. Development of construction planning documents from the construction discipline requires inputs and comments from engineering and other disciplines but the planning work process is basically a WE for the construction discipline.

We find that organization of information in these two levels greatly reduces the confusion on what is the right information to use and when to use it. Information is managed with more efficiency, better organization, and integrity.

## 2.5 Document Management

Document management in a library serves the basic functions of storing, tracking, retrieving, and distributing of documents. To facilitate tracking, documents are classified as different types with listing of the document title, numbers, revisions, dates, and disciplines, and so forth.

For electronic documents, the management functions are much more efficient because documents are tracked as files in computers with different file formats (such as DGN for CAD drawings, DOC for electronic documents, PDF for format-neutral documents, and so forth). Documents are simultaneously identified and organized in databases that can be retrieved for viewing or distributed efficiently through networks to multiple users. Documents can further be linked to other documents through databases.

Content management in documents is an advanced level of document information management. Information can be extracted from intelligent documents such as CAD drawings or from word processing programs. Information extracted from documents can be listed and organized as databases for other applications.

There are different types of document management programs—some for general purposes, some for technical documents in a working environment, and some for documents in a controlled environment with rigid control for viewing, retrieving, and changes. There are standard document management features, including document check-in and check-out control, features to accept various document formats, viewing of documents through network, and long-term storage management.

Workflow management is another feature of document management. The system can set up for auto-distribution or alerts, action item tracking, document review and approval tracking, and so forth. A good document management system is a flexible and efficient tool for report generation. Templates can be developed to generate various document summary reports for delivery to client, to indicate mechanical completion, or to document project close out.

A document management system is one of the main tools for information management because today documents are still the dominant information type.

## 2.6 Information Warehouse

As information is generated from various CAD and application programs in the form of drawings, specifications, databases, and 3-D models, Figure 2.1 shows how the information is typically exchanged from the provider of the information to the user of the information. As shown, it is very complicated that user E must go to source A, B, C, and D for information while user F, G, and H similarly must go to the same source A, B, C and D for information. There may be duplicated and out-dated information in these sources and additional efforts are required to manage this information.

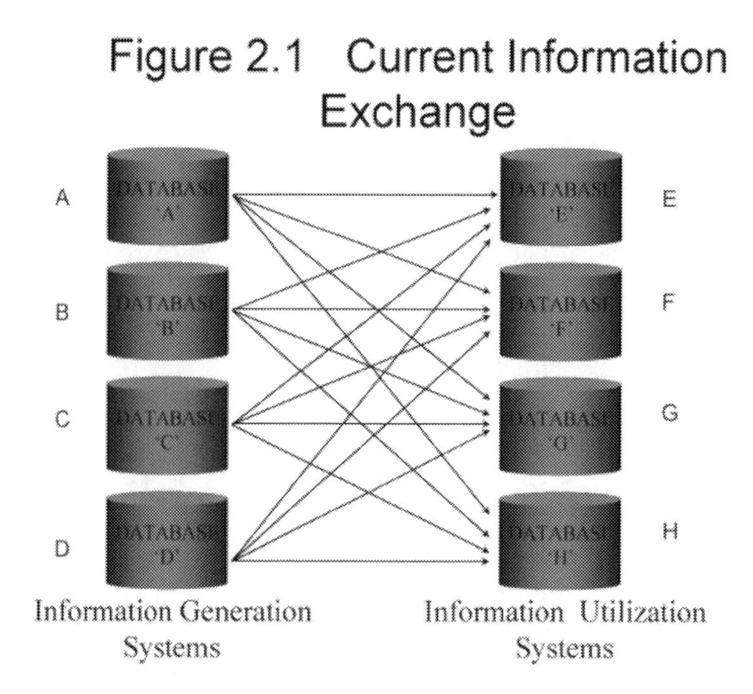


Figure 2.1 Current Information Exchange

The preferred approach is to exchange information from the users through a single source as shown in Figure 2.2 that we call it a single source information data warehouse. In this approach, information users E,F,G, and G all retrieve the same information from the information data warehouse provided by the same information sources A,B,C and D

Figure 2.2 Information Exchange through Information Data Warehouse

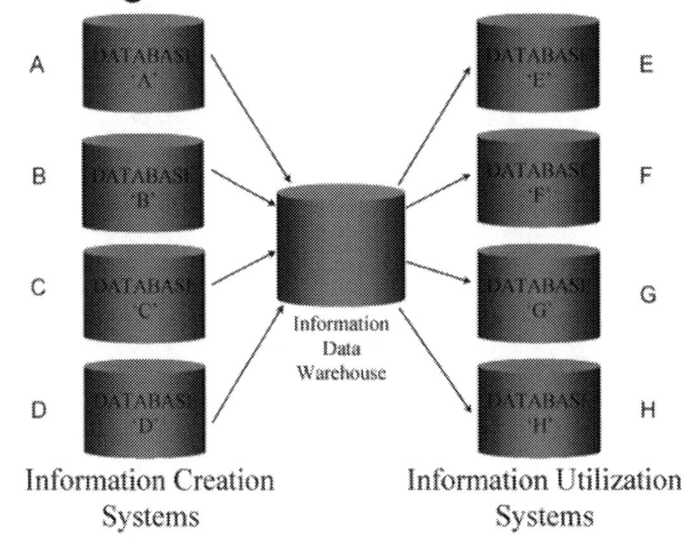


Depending on the specific type of plant and the project organization, different information can be stored in the information data warehouse, but there are some basic systems and operational requirements. The system requirements include the ability to link documents to documents, documents to data, and data to data. The operational requirements include being Web based and having features to receive, store, retrieve, view, publish, and transfer information efficiently. Long-term, historical record also must be storable.

Information stored in the information warehouse is usually in ICE, so it provides single-sourced controlled information with guaranteed integrity. It has the ability to link the three basic types of information, documents, databases, and 3-D models. In a project, information may be stored directly in the information data warehouse or, as shown in Figure 2.3, each information source may be stored in its own database program and server that is linked to the information data warehouse server.

# Figure 2.3 Information Data Warehouse Linkages

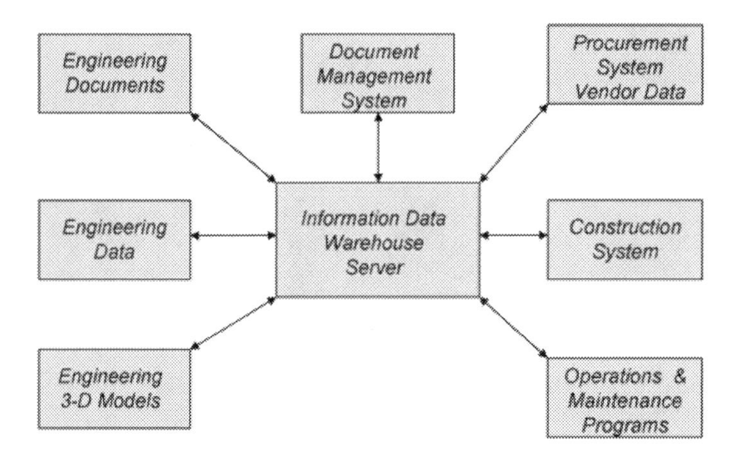


The information warehouse is for the plant lifecycle that starts from engineering and design and expands to the procurement stage with vendor information, then to construction, and finally to operations and maintenance. With each step, the warehouse may be restructured to fit the information content and format requirements. At the end of the project EPC phase, there may be a major regroup of the warehouse information for turn-over. More emphasis is on process-related information such as P&IDs, process equipment, and instrumentation on safety rather than on facility support items, because such items more concern operation of the plant while vendor information on equipment and components concerns maintenance.

## 2.7 Integration and Automation

In the plant lifecycle, information flows from engineering and design to procurement, to construction, to operations and maintenance. In each step, there is a linkage of input information required from the upstream discipline and then a linkage of output information to the downstream disciplines. Integration is linkage and management of these information relationships.

Once the input and output relationships are established, information integration provides information integrity through single-source information to minimize efforts in information searching and to give better coordination efficiency. There are different methods to integrate the three basic information types. Tag as an intelligent object has attribute database that is identified in documents and 3-D models. In a process plant, extensive usage of tags is an efficient approach to link information.

Automation is the performance of certain work processes with no or a minimum human interfaces. While automation increases productivity, it has to satisfy two basic requirements. First, there can be no automation without integration or linkage of information. Second, rules must be developed to define input and output information requirements to control the automation process correctly.

Historically, the first successful application of automation was in civil engineering where site contours were generated automatically from site surveyed data. In a process plant, 3-D design became useful only after the successful development of automatic extraction of piping isometrics from piping 3-D models.

Even with advancements in computers, database applications, and software development, one must be careful about automation. It is often difficult to define proper automation process rules to cover different situations. It is also necessary to weigh the development costs versus the potential savings of automation. Sometimes, insertion of human decision at the critical step may simplify automation steps, improve efficiency, and save cost

As we proceed, we'll see that more and more integration or linkage of information is the key successful element in the management of plant lifecycle information.

# CHAPTER 3
# ENGINEERING

## 3.1 Overview

Through process requirements from the owner, front-end engineering translates process information into a set of engineering design information that provides inputs to detailed engineering and determines the cost and performance of the plant. Engineering also generates all of the design information that procurement and construction will use to build the plant.

In the detailed engineering, work is divided into two groups: one is process-related and one is facility or configuration-related. The process-related disciplines include process, mechanical processes (P&IDs), instrumentation, piping, and process equipment, supported by electrical. Information generated from these disciplines not only feeds into design of the configuration and facility but also is an important part that controls the production of the process products and manages the operations and maintenance of the plant.

Facility-related disciplines by contrast include architectural, civil and structural, mechanical, piping, electrical, and facility equipment. These disciplines generate information so procurement and construction can physically build the plant. Once the plant is built, most of the facility-related information has no direct use for the daily operation of the plant except when there are changes or major modifications to the plant.

Detailed design engineering information is tracked through basic information types of documents and databases that can be represented by the extensive use of tags. The third major information type, the 3-D model, is used for configuration coordination and management and is also tracked and linked with documents and databases.

As stated earlier, information management means linking information both from discipline to discipline and also from step to step in each work process. More and more application programs have been developed to link information. By such integration, rules have been developed for automation applications. There has been good intra-discipline success with integration through computer-integrated engineering (CIE). If one draws a boundary around an engineering discipline with a clear definition of input and output information across the boundary, then it is easier to integrate and automate information and applications within the boundary of the discipline.

Inter-discipline engineering integration has proved more difficult because it involves coordination and information flow from several different engineering disciplines. Depending on the specific type of process plant, there are different work processes for each discipline. With the rapid development of computer, server, and database technologies in recent years, engineering information data warehouses have provided a practical and efficient approach for achieving a truly single-source information and for promoting better information exchange, coordination, and integration. An example of an engineering information and data model set-up for an information data warehouse is shown in Section 3.8.

## 3.2 Process and Front-End Engineering

Process in a process plant determines first the function and hence the product of a plant. It can be a chemical product, a food or pharmaceutical product, or a product as represented by the generation of electricity or clean water. It is usually the owner's responsibility to select and define the process for the product. The process can be a copy or an improvement of an existing process, or a first time application of a new process from a pilot plant. The owner also specifies the required output quantity of the product, which determines the size of the process plant. The process information generated from the process analysis includes all the data that governs the performance of the plant. Such data through front-end engineering are then converted into engineering information for the detailed engineering design. This is usually represented by process flow diagrams and the corresponding stream functions, process control approaches, process equipment and material selection diagrams that define the material requirements for the process.

At this stage, the owner or his selected EPC performs preliminary engineering to determine the basic plant parameters and to generate sufficient information for a preliminary cost estimate of the plant. This information—in combination with the potential sales price of the process product and the market economy—generates financial information for deciding whether to proceed with the process plant project. The execution of this part relies heavily on the owner's previous knowledge and experience on this type or similar type of process plant.

Front-end engineering is a more formal way to translate process information into the engineering information that is required not only for detailed engineering but also bounds the cost of the plant. Typically, it includes P&IDs with instrumentation, safety and defined process control approaches, major process equipment specifications, and the physical size of the plant, including plot plans as determined by the preliminary piping layout. For a process within a building, the preliminary size of the building and its supporting structure are also defined. Currently, there are programs that can automate some or part of the front-end engineering for certain kinds of process plants. The problem is that this type of program cannot always handle different types of processes because it cannot always incorporate information from many commercially available process equipment. Furthermore, a plant layout as generated by the automatic program cannot always efficiently fit into the available plant site.

As front-end engineering is the first major step in the development of a process plant, information as generated is very critical for applications throughout the lifecycle of the plant. At this stage, it is not too early for procurement, construction, and operations and maintenance to participate in front-end engineering. Their inputs have deep impacts on the performance and financial success of the plant.

Due to budget and other considerations, there is sometimes insufficient effort spent during the front-end engineering looking at trade-offs of alternative designs and physical arrangements of the plant. As the project proceeds to detailed design, design changes can be minimized if all the alternative designs and their costs have already been studied at the front-end engineering stage. This can provide major cost savings.

A good front-end engineering sets 90 percent of plant costs and provides information for detailed engineering and the subsequent plant lifecycle steps. The goal is to generate the required information efficiently at the front-end so changes are minimized in the subsequent steps.

## 3.3 Detailed Engineering - Process-Related Disciplines

As mentioned earlier, we want to consider detailed engineering in two parts: first in terms of process-related disciplines and then in terms of facility or configuration-related disciplines. Figure 3.1 shows the information relationship among the five basic process-related disciplines: process, mechanical process (P&ID), process equipment, instrumentation, and piping plus electrical support.

# Figure 3.1 Process Related Disciplines Information Relationship

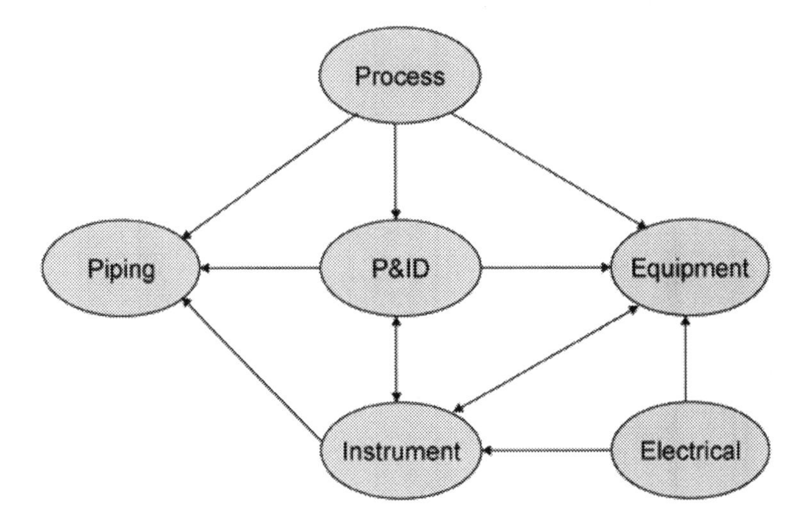


The mechanical process transforms process information and stream functions from the process discipline into engineering information represented by P&IDs. The process discipline also defines process equipment information and requirements that are incorporated into P&IDs as well. Instrumentation is a part of P&ID also because it sets process measurements, safety, and control requirements. Instruments and control systems as shown in modern P&IDs are much more complex because instrumentation is no longer mechanical, but is virtually all electrical. Control systems are all computerized, digital, and inter-related to satisfy strict safety and environmental requirements. Electrical serves a support function by providing the required power for process equipment and instrumentation and control system.

The piping discipline is considered as a part of the process discipline because schematic pipelines are process lines in P&IDs and because they connect process equipment and control system items such as valves. Piping dictates the arrangement of process equipment according to flow, heat expansion, and operations and maintenance requirements. Thus, piping sets the basic configuration and site size of the process plant.

As one can see, P&ID is the key that connects the process-related disciplines. It provides information on the process line summary, the valve summary, the instrument summary, and the process equipment summary. Figure 3.2 shows the information flow among the process-related disciplines. As mentioned earlier, the preferred approach of information management is for the discipline to post and retrieve information from a centralized, single source as in an information data warehouse.

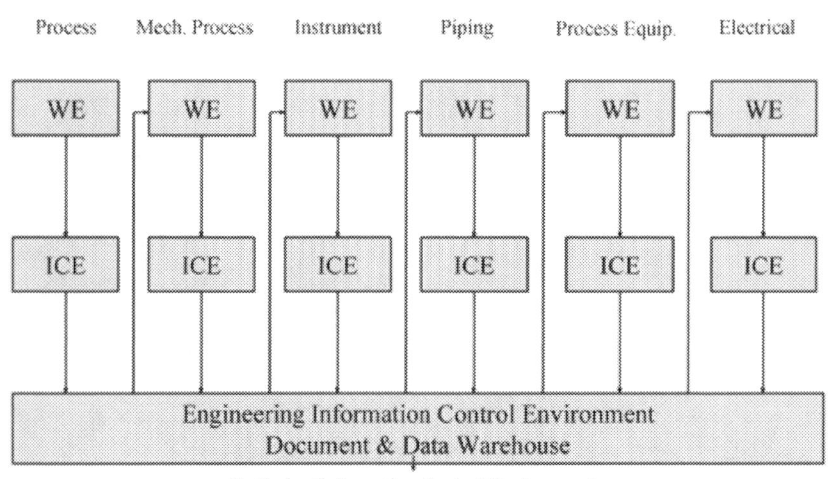


By this scheme, controlled information (ICE) from other disciplines and required outside information are retrieved as inputs and used by the discipline to do the work. Each discipline performs various tasks within the discipline boundary that are mostly in the discipline's WE. This means such information as generated in the WE by the discipline is subject to change on a daily or hourly basis. Coordination of such changes is easier to manage as they are contained within the discipline. If an application program is used for the tasks, information flows and is contained within that application program's database.

When appropriate, some information from the discipline is organized and posted to the information warehouse as controlled information in ICE to be used as inputs by other disciplines. Again, once information is posted in a controlled environment, it can only be changed by the discipline that provided the information. Controlled information changes can be posted by notification or on an agreed schedule.

Following is an example of information flow on developing P&IDs. Mechanical process discipline retrieves controlled process information from the process discipline. It then develops P&IDs from this information. At the beginning of the detailed design, there are considerable changes, so preliminary P&ID's are posted weekly as controlled information to the information data warehouse. The other process related disciplines retrieve and use the posted P&ID information with confidence because all disciplines use the same controlled information with guaranteed information integrity. As detailed engineering proceeds, less and less changes occur and new controlled P&ID's are posted less often only on a required basis.

Experience shows that efficient control of information flow is the key to design success. The WE and ICE concepts are just one approach for managing the what and when of information flow. Other terminologies and methods can be used, but the main point is how to efficiently achieve information integrity from a single source with accuracy and currency. This approach does not impose any restriction on the flow of information. Disciplines are free to discuss design approaches and to exchange preliminary information as long as disciplines are aware that at the various stages of design, only controlled information from other disciplines should be used as inputs for their work to produce outputs.

## 3.4 Information Management and Integration through Documents and Tags

Before the development of computers and CAD, paper documents were the only means for presenting and communicating information. In a typical process plant, there is a cascade of many drawings representing the flow of information from discipline to discipline. This is supported by documents like specifications and datasheets. Table 3.1 shows a typical list of process-related documents.

Table 3.1 Typical Process Related Documents

Process Related documents

Line Summary (Process Line Tags)
Valve Summary (Valve Tags)
    Data Sheets
    Specifications
Instrument Summary (Instrument Tags)
    Data Sheets
    Specifications
Process Equipment Summary (Equipment Tags)
    Data Sheets
    Specifications
Piping Specifications

Process Related Drawings

Process Flow Diagrams
Materials Diagrams
P&ID's
Electrical Single Lines
Electrical Interconnect Diagrams
Instrument Digital Control Diagrams
Instrument Loop Diagrams

There are two basic problems with managing information from documents. First, because documents are subject to interpretation, it is uncertain if other disciplines correctly understand and retrieve the right information from a document. Second, as information flows from documents to document how can we be assured that we maintain information integrity (accuracy and currency) from one step to the other step. Historically, information management of documents relied heavily on work-force experience and the disciplines working as a team with extensive checking of documents.

With CAD and other computer-generated documents, not only is generation and development of documents more efficient, but also documents have become 'intelligent.' Computer-generated documents can be tracked and linked electronically. Information content within the documents can be identified, retrieved, tracked, linked, and organized as databases. While engineering information today is still represented mostly by documents, the increasingly preferred approach is to show information as data. In theory, engineering documents can basically all be generated from databases.

The term 'tag' has been used extensively in many engineering applications and documents. We would like to define tag as an object that contains attributes represented by a database. Tag is also an engineering term that represents an item in a document or an item at the location of a design. The following illustrates how the tag works.

First, we look at the process line in a P&ID. If we call a process line number as a process tag number, then it contains a database. Table 3.2 shows a typical list of attribute data for a process tag number. Assembling a group of process tag numbers in a spreadsheet constitutes a typical process line (tag) list. If we change an attribute in a tag, it changes the tag's database and changes the process line list.

Table 3.2 Typical Attributes of a Process Line Tag

| Column | Description |
|---|---|
| Line No. | The series of letters, numbers, and size that uniquely identifies each line |
| P&ID No. | Piping and Instrumentation Diagram (P&ID) in which the line originates or terminates |
| P&ID Rev | The revision of the P&ID |
| BLDG | The building/area in which the line is located. |
| From | The connecting line (tag) or equipment (tag) where the line originates |
| To | The destination line (tag) or equipment (tag) where the line terminates |
| Phase | Phase of the containing fluid as indicated by L=Liquid; V=Vapor; M=Mixed Phase; L/V=Liquid and Vapor |
| Oper Press (psig) | Operating pressure based on process flow diagram stream data and line hydraulics |
| Oper Temp (F) | Operating temperature based on process flow diagram stream data |
| Design Press (psig) | Design pressure, based on Basis of Design |
| Design Temp (F) | Design temperature based on Basis of Design |
| Hydro Req'd | Hydrostatic testing required? Y=Yes; N=No |
| Insul. Thk (in.) | Insulation thickness of lines |
| Insul. Mat'l (in.) | Insulation material of lines |
| Paint Code | Paint code per coating requirements |
| Qual Level | The Quality Level designation (Q or Non-Q) |

In addition to a process line tag, there are valve tags, instrument tags, and process equipment tags. By definition, all these tags are objects that contain attributes in databases. Table 3.3 shows typical attributes for a valve tag, Table 3.4 for an instrument tag, and Table 3.5 for a process equipment tag. Grouping each of these, they become a valve summary list, an instrument summary list, and a part of the master equipment list, respectively. Furthermore, in a tag database, it can contain information that links the tag to another documents or a 3-D model.

Table 3.3 Typical attributes of a Valve Tag

| Column | Description |
|---|---|
| Tag No. | The series of letters and numbers that uniquely identifies each valve |
| P&ID No. | Piping and Instrumentation Diagram (P&ID) in which the valve is located |
| Valve type | The valve type |
| Valve Size | The valve size |
| Body Material | The body material |
| Trim Material | The trim material |
| Line No. (Tag) | The series of letters, numbers, size, and pipe class that uniquely identifies each line |
| Room No. | Identify the room number the valve is located |
| P&ID Loc. | Identify the location of the vavle on the P&ID |
| Qual. Level | The Quality Level designation (Q or Non-Q) |

Table 3.4 Typical Attributes of an Instrument Tag

| Column | Description |
|---|---|
| Tag Number | The series of letters and numbers that uniquely identifies the instrument |
| Description | Description of the instrument |
| Service | The service or usage of the instrument |
| P&ID No. | Piping and Instrumentation Diagram (P&ID) in which the instrument is located |
| Line or Equip. No. | Line or equipment number in which the instrument is located |
| Mounting Detail | The mounting detail drawing number that shows the instrument is installed |
| Control System | The control system that shows the instrument |
| Instrument Plan | The instrument plan that shows the instrument |
| Piping Plan | The piping plan that shows the instrument |
| Loop Diagram | The loop diagram that shows the instrument |
| Logic Diagram | The logic diagram that shows the instrument |

Table 3.5 Typical Attributes of a Process Equipment Tag

| Column | Description |
| --- | --- |
| Tag Number | The series of letters and numbers that uniquely identifies the equipment |
| Description | Name or description of the equipment |
| Building/Room No. | Identifies the building/room number where the equipment is located |
| P&ID Number | The Piping and Instrumentation Diagram (P&ID) in which the equipment is shown |
| PFD Number | The Process Flow Diagram where the equipment is shown |
| HP | Estimated motor nameplate horsepower for the given piece of equipment |
| KW | Estimated motor nameplate kilowatts for the given piece of equipment |
| Drive Type | Type of electric equipment drive |
| Load Class | X= Automatically reenergized on the emergency generators, X-Manual= Manually reenergized on the emergency generators per operator discretion |
| Active Standby | Any equipment that is either actively running or a spare |
| Quality Level | The quality Level designation (Q, Non-Q) |

Note that all of these tags are from P&IDs. This is why P&ID controls and links the process-related disciplines. Once the tags are defined with attributes, they are intelligent objects and can be tracked and linked with other tags and documents like drawings and datasheets. Throughout the lifecycle of a process plant, the process tag information is used over and over again and is the key element for control of the plant in operations and maintenance.

## 3.5 Detailed Engineering - Facility and Configuration Related Disciplines

From information generated by the process-related disciplines and stored in a single-source information data warehouse, the physical plant is then designed by the configuration and facility-related disciplines. As shown is Figure 3.3, these disciplines are piping, equipment, architecture, structural and civil, mechanical, and electrical.

## Figure 3.3 Configuration and Facility Related Disciplines - Information Relationship

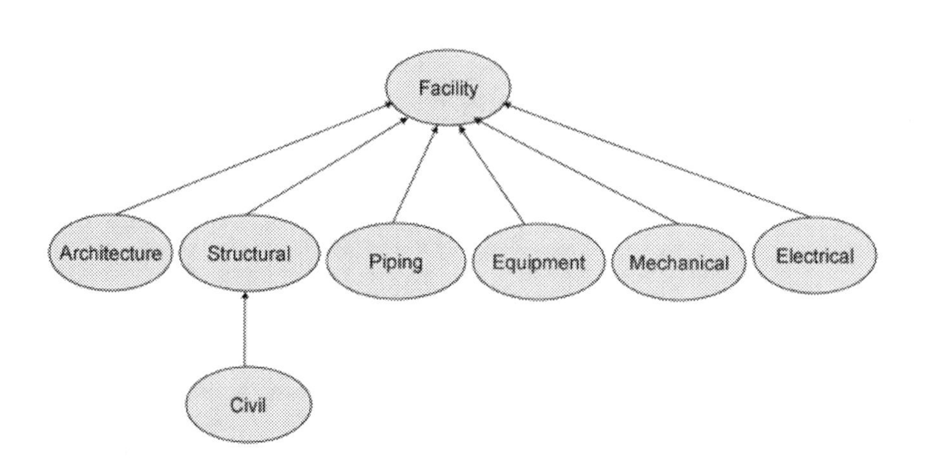


As already noted, piping is part of the process disciplines because it contains the flow of process media and connects to process equipment, control valves, and instrumentation. Piping is also a part of the configuration disciplines because it not only has physical size but also governs spacing of the major process equipment for maintenance and operations. Thus, piping designs the plot plan and determines the physical size of the plant.

In many process plants such as those in the pharmaceutical and food industries, process functions and equipment are enclosed in a building. In that case, architectural is the major discipline that designs not only control rooms and offices but also major facilities and buildings to enclose the process piping and equipment.

The structural team designs facilities such as pipe ways to support pipes, foundations for equipment, and structural elements for buildings. This team also designs large concrete tanks and pools to serve part of the process functions for water and wastewater treatment plants. The civil team develops the plant site, designs the road network inside the plant, connects utilities from plant to the outside battery limit, and supports the structural team's design.

Inside a building, the mechanical team designs HVAC systems, including ducts and heating and cooling equipment, elevator and cranes, and basic utilities. In some plants, fire protection is considered part of the mechanical system. In a nuclear facility, mechanical provides cascade airflow, special HEPA filters, and so forth for the critical element of ventilation. In a mining process plant, mechanical equipment such as conveyor belt systems, crushers, and so forth are part of the mechanical process line.

Electrical provides power for process and mechanical equipment, instrumentation, and electrical utilities. Electrical equipment includes motors, generators, transformers, and motor control centers. Routing of cable trays and conduits occupy also physical spaces. Modern digital instruments and control systems require computers and special items that are related closely to electrical.

In addition to process equipment, there is electrical equipment and mechanical equipment as well. Equipment are all tagged items that contain attribute information in a database similar to the process disciplines. Assembly of all equipment items forms the master equipment list, which can be an intelligent document represented by databases in a spreadsheet.

Figure 3.4 shows the information flow between the configuration and facility-related disciplines. Each discipline generates design information within its working environment. It gets input information from the same central engineering information data warehouse that contains controlled information from the process-related and other facility-related disciplines.

**Figure 3.4 Configuration & Facility Related Disciplines – Information Flow**

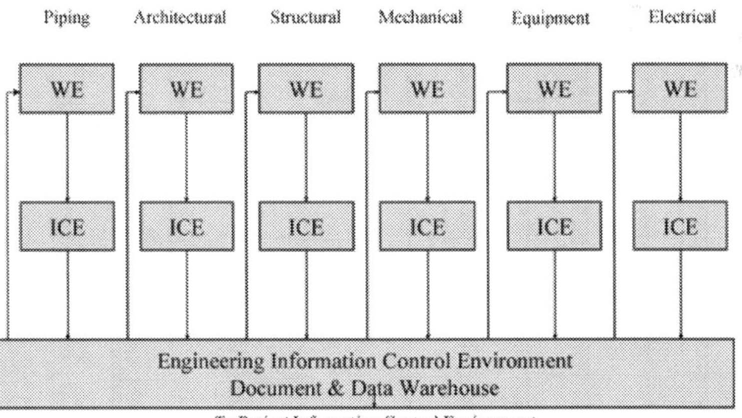


Traditional drawings are still used to show the designs from configuration and facility-related disciplines. However, CAD-generated drawings include intelligent databases that can be linked to other documents such as other drawings, specifications and datasheets. Tags are used as objects with databases to represent electrical and mechanical equipment and items. Piping isometrics can also have tag numbers that link directly to process line tag numbers. Architectural items such as doors and windows and structural steel components may be tagged, but the traditional use of piece marks to track these items may be sufficient for procurement and construction.

In an open plant, the plot plan determines the size and configuration of the plant. In a closed plant, architectural designs the size and configuration of the building keeping in mind the process requirements. In the lifecycle of a plant, design information generated from configuration and facility-related disciplines relates directly to procurement and construction of the plant. Once the plant is built, information from most of these disciplines may have limited direct usage in operations but is important for plant compliances and changes.

## 3.6 Control of Physical Configuration Through 3-D Models

Advances in computer-generated 3-D design of process plant have considerably improved accuracy and provides efficient management of a plant's physical configuration. First, we want to emphasis that 3-D model is the full size electronic representation of the physical plant. Thus, there is no scale factor in a 3-D model. When extracting a drawing from the model, a scale may be picked so the drawing fits the given drawing size. In one sense, we may want to call 3-D model a '3-D full size model'.

If we understand this concept, then the 3-D design of a plant represented by a 3-D model is an assembly of the physical components of the plant. Figure 3.5 shows the disciplines that do 3-D design. Information from each discipline is consolidated into a composite 3-D model. Figure 3.6 shows a typical 3-D model of a process plant.

## Figure 3.5 Configuration Management and Control through 3-D Model

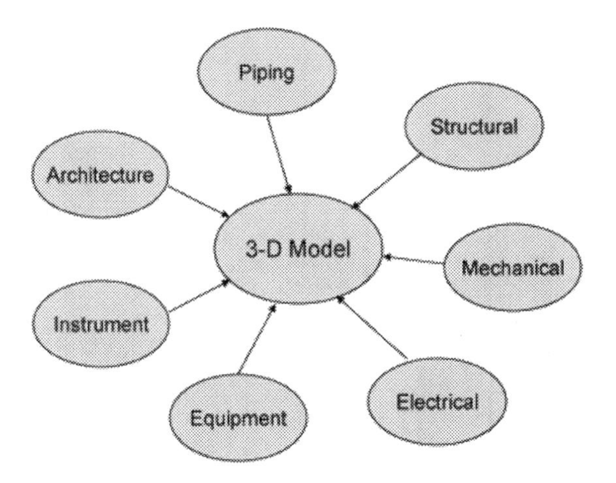


During design, one discipline—usually piping or architecture—is assigned responsibility for coordinating the configuration or usage of space in a 3-D model. Each plant component occupies its space and if the design is correct, the assembly of components should all fit together with no interferences. In fact, one of the key advantages of 3-D model design is the ability to do automatic interference detection. This can be a major cost saving feature because there then are minimum changes in the field due to interferences.

The 3-D model also becomes the central location for design information coordination. Dimensional information can be obtained by measuring the model. Touching the model components with tags opens the database information associated with the tags and connects to related documents. Linkage information of the components is also shown in the model. Additionally, the 3-D model gives information on accurate size measurement and bulk material take-off of items such as cable trays, conduits, and ventilation ducts.

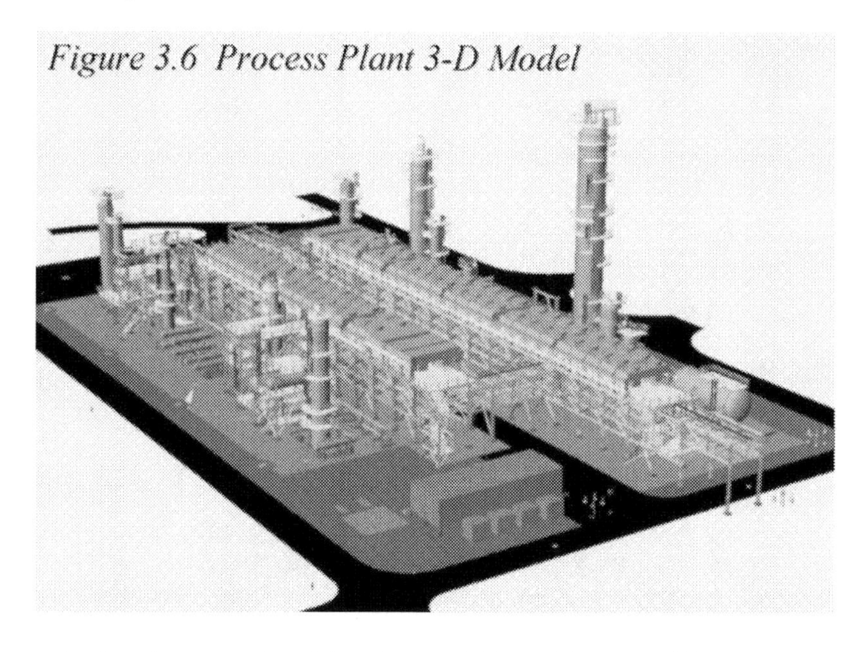
*Figure 3.6 Process Plant 3-D Model*

In the subsequent plant lifecycle steps, 3-D models are used for construction planning, construction sequencing, construction status reporting, and so forth. In the plant operations, models are used for operator training, safety evaluation, operational trouble shooting, and other tasks.

## 3.7 Computer Integrated Engineering (CIE)

Engineering work involves design from drawings to 3-D models, to engineering and analyses and to materials take-off from bulk to equipment and component items. If one draws a boundary around one engineering discipline as shown in Figure 3.7, then within the boundary, information feeds from one work process to the other until each step of work is completed. Controlled information is inputted across the boundary to the discipline from client requirements, design criteria, outputs from other disciplines, and other outside sources. The input information requirements are defined so as to minimize information searching, using wrong information and unnecessary changes that may increase costs and delay schedule. On the other hand, the discipline's controlled output information is also defined and published across the boundary to satisfy the requirements of other disciplines. As discussed earlier in Figures 3.2 and 3.4, this input and output of information through the discipline boundary is retrieved from and loaded to a single-source information data warehouse.

Figure 3.7 CIE Intra-discipline Integration

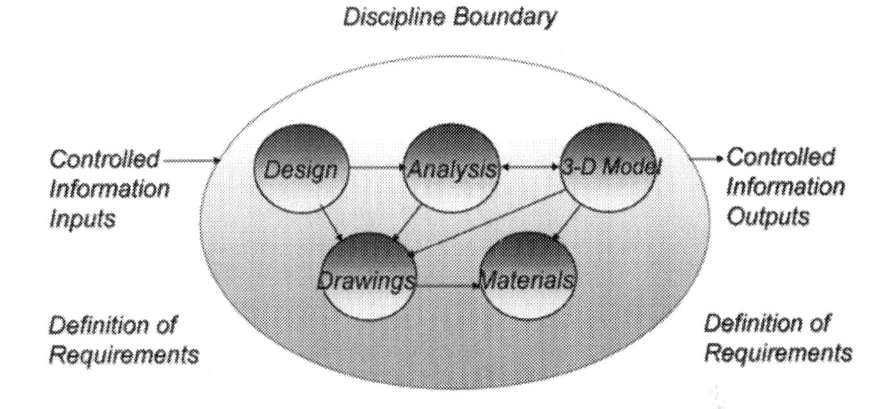


Within the discipline, we want to integrate or to link information flow from each step in the work process. We call this approach computer integrated engineering (CIE) for intra-discipline applications. This concept is not new because integration has already provided a basis for automation. However, with advances in computer and database technology, many powerful engineering and design application programs have been developed. The efficient use of such programs requires not only understanding of the technology but also changing the work process. This change is not always obvious because an iteration process is sometimes required before a new work process is developed to fit the new technology.

Following is an example of intra-discipline integration for piping design. Piping uses a piping design program to develop a 3-D piping model. Piping specifications and piping components are pre-loaded in the program. During the design, piping dimensional and size information from the 3-D model is transferred to another program for analyzing piping flexibility and stress. Piping isometrics are then extracted from the 3-D model along with a proper presentation of all the pertinent information. If necessary, piping drawings can be extracted and produced from the 3-D model. As one can see, all of these work processes stay within the piping discipline boundary while controlled output information such as piping drawings, isometrics, 3-D model, and so forth are published across the boundary to an information data warehouse as ICE for use by other disciplines.

Another example is the structural design of a pipe way support structure. A 3-D structure model is first developed through a structural program and then stress analysis is done from a finite element program. Loadings from the pipe supports are fed into a concrete footing design program. An automation program can be developed to further integrate these steps of the work process. The difficulty lies in developing rules that govern what information is required and linked at each step of the work process. For example, the finite element program generates separate loading for each support. Should we use maximum loading to design all footings or should we put loadings into two or three groups for the design of two or three different types of footings? Would a manual decision at this point be more efficient than rules for automation?

If one draws a boundary to enclose all engineering disciplines as shown in figure 3.8, this represents Computer Integrated Engineering (CIE) for inter-discipline integration. Again, within the boundary, there is a flow of information from discipline to discipline. To advance the design, each discipline's working environment generates information that is linked through a communication and application program from discipline to discipline. However, because information in a working environment changes on an hourly or daily basis, to ensure information integrity from each discipline published through an engineering boundary only controlled information must be used.

*Figure 3.8 CIE Inter-discipline Integration*

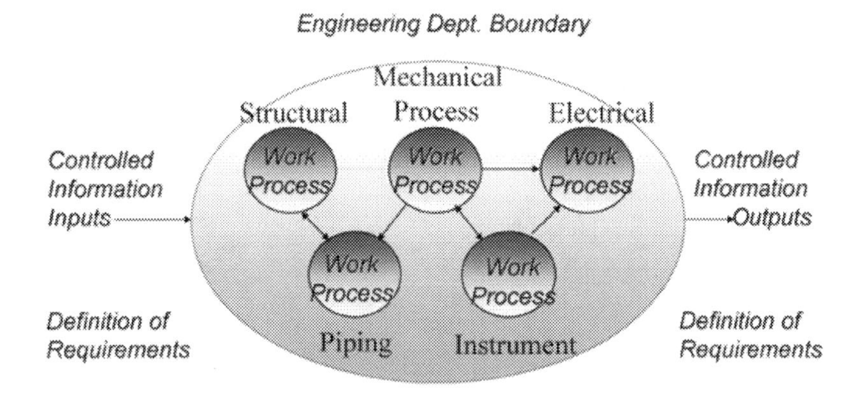


Inter-discipline integration and automation are more difficult because there are interactions of design information requirements from discipline to discipline that are not always available in a timely manner. For example, the structural team has to estimate the pipe loading on the pipe way design before the piping team can finish the design and analysis of pipe lines. Similarly, the electrical team has to estimate the horsepower requirements for the electrical design before the process equipment design can be finalized. To complete the process equipment specification, piping needs to provide nozzle loading. The logical and efficient progression of multi-discipline design requires information coordination and management, which is best accomplished with information flow through a single-source information data warehouse. Nevertheless, a 3-D model serves as the key element for inter-discipline coordination and integration of the physical configuration of a plant.

Controlled information outputs from the engineering disciplines are published across the engineering boundary to an information data warehouse. Documents, databases, and 3-D models are organized to be retrieved and used efficiently with confidence by the downstream disciplines in the plant lifecycle.

## 3.8 Engineering Information Data Warehouse

An information data warehouse has been mentioned several times as a useful element in lifecycle information management. How one constructs an engineering information data warehouse is now examined.

Starting with the three basic information types—documents, tags (databases), and 3-D models (see Figure 3.9)—these are connected in the information data warehouse. For example, a valve tag is shown in a P&ID document, on a valve summary and valve datasheet, and also as an item in the 3-D model. A document and its related file can be stored in the data warehouse or in a separate document management system that is connected to the data warehouse via databases. Document attribute information—document number, revision, date, and so forth—are used to identify the connection. The 3-D model and its related information are also stored in the data warehouse or in a separate 3-D program that is connected to the data warehouse via databases.

**Figure 3.9 Information Data Warehouse**
**Three Basic Information Blocks**

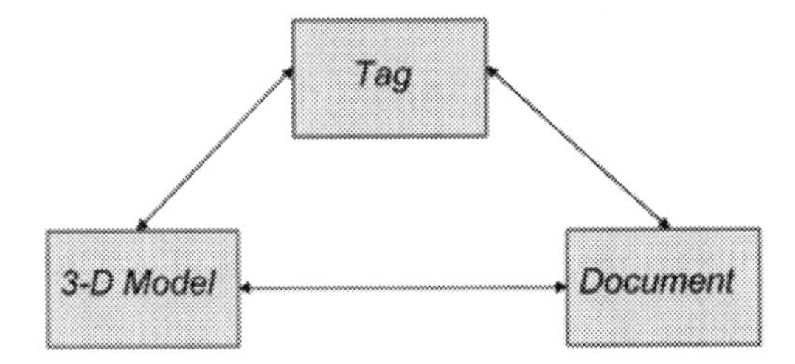

To track information more efficiently, a hierarchical setup of information blocks such as site area, design area, system, and discipline are added to the data warehouse (see Figure 3.10). These blocks are then connected to the basic information types. For example, the mechanical process discipline can track a P&ID in a high-pressure system in the relevant design area. Similarly, an instrument tag located in a 3-D model can be tracked through safety control system in the special design area.

Figure 3.10 Information Data Warehouse Information Blocks

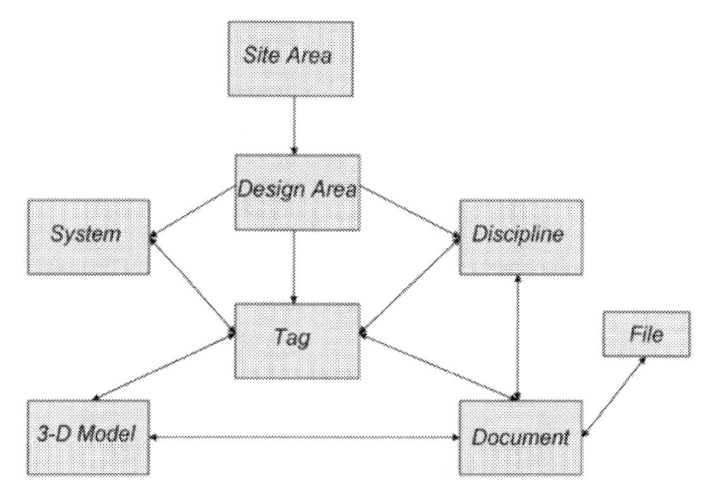


By definition, all information stored in the information data warehouse is controlled information. This means only the originator can change the information via a controlled change process. To accomplish this, documents are connected to their files and must be converted into a format that maintains its intelligence for linkage while remaining locked from changes. (This will be format dependant. In AutoCAD, for instance, the DWG format converts to a DWF format). Some documents are loaded to the data warehouse in the PDF format or in their native format but locked for reference usage only. P&IDs, for instance, are published in the PDF format for viewing and use by the process-related disciplines and the client. Piping plot plans can be published in the CAD DGN format for reference use only. Equipment datasheets are in PDF format as part of the procurement package for bidding by equipment vendors. For some specific requirements, documents may also be linked to files in their native format as in the working environment. In this case, controlled document check-in and checkout are required to maintain the document's information integrity.

A tag is shown as one main information block in the data warehouse because, in engineering, tags with the attribute databases are the main information linkages to documents and 3-D models. Spreadsheets provide the common approach for inputs of tag databases. For plant design, there are programs with different approaches for generating 3-D models. Some programming efforts are required to link databases from an information data warehouse to 3-D model programs for viewing and extracting component information.

The above describe basic requirements for setting up a good information data warehouse. It should also be Web based so users can communicate with it effectively. Properly set up, it can store, track, retrieve, view, publish, and transfer information efficiently. It has the flexibility to link and interface with engineering application programs and other programs like document management, procurement, project controls, and so forth. It keeps and tracks historical and long-term records. An information data warehouse does not delete any information without special actions and stores and tracks all changes. The importance of this is discussed further in the next chapter on Project Compliance.

# CHAPTER 4
# PROJECT COMPLIANCE

## 4.1 Overview

Project compliance does not generate information as in engineering. Rather, it extracts and gathers information to show that the plant as designed and engineered meets the owner's specifications. Procurement has to satisfy that all bulk materials and tagged items are purchased as specified by engineering. Construction has to go through extensive checking and testing to verify that the plant is built as designed. Traditional QA (quality assurance) and QC (quality control) are also part of project compliance.

To satisfy plant safety and to comply with environmental, governmental, and regulatory agency regulations are also a major part of project execution toward fulfilling project compliances. Depending on the type of process plant, project compliance can be a major cost item (e.g., pharmaceutical or nuclear process plants). Failure on this part can have disastrous impact on schedule and cost.

As the plant moves through each step of its lifecycle, it is inevitable that there are changes in engineering and design, in procurement, and in construction. Thus, it is important to track and document changes, control plant configuration, and develop change-management procedures. These are all parts of project compliance.

In project compliance, information management deals with verification and record keeping of information. The key here is still to determine what information needs to be managed. To ensure success, compliance requirements and proper procedures are established at the beginning of the project. Long-term information storage, including the ability to efficiently retrieve historical information, is an added feature for project compliance. A good information data warehouse can satisfy this requirement.

## 4.2 Quality Assurance and Quality Control (QA/QC)

The QA/QC organizations perform most of the traditional functions of project compliance in a project. At the beginning of the project, it is important to establish QA/QC requirements, including organization, responsibility, and procedures. Depending on the type of process plant, there can be different levels of control and complexity. But, the main task is still to collect, verify, and record information so that the intended lifecycle functions of the plant are satisfied.

As stated earlier, information management links different types of information at each step of plant lifecycle work processes. With linkage and integration of information, automation of work process is developed. This means that the traditional checking of information that normally occurs at the transitions for each work process step is greatly reduced because information integrity is now guaranteed by the single-source, controlled information source as in an information data warehouse. Reduction of QC efforts is one of the major cost-saving benefits that information management provides.

However, the reduction of QC entails a corresponding increase of QA. The initial setup of information input and output requirements needs to be developed. Design application programs and automation procedures need to be verified so that the outputs are correct. Information linkages in an information data warehouse produce the intended correct results. In other words, we want assurance that information integrity is indeed achieved throughout the lifecycle work processes.

Once this type of QA is established for one type of process plant, it can then be applied to many similar plants thereby providing a major reduction of QC effort. The overall net effect is a potential reduction of total QA/QC cost.

## 4.3 Safety and Environment Compliance

In some process plants, satisfying safety and environmental requirements is second only to the satisfactory production of products. Safety and environmental compliance are required throughout the lifecycle of the plant. Failure to be in compliance can delay start-up and/or incur very costly remedial actions. In some cases, non-compliance may result in the plant's operating license not being issued.

In engineering, safety and environmental compliance are integral part of the design requirements as defined in design criteria or design bases. For a refinery or chemical plant, there are regular hazard reviews. For nuclear power plants, formal safety and hazard analyses are required, which can include qualitative failure mode and effect analyses and even quantitative analysis to determine the probability of failures.

In procurement, all purchased equipment and components need to satisfy the same safety requirements as specified in the criteria. In construction, safety relates more to the construction processes to protect workers. Safety plans and procedures are required throughout the construction. On the other hand, after construction, extensive tests are required to verify that the systems are built and can be operated safely.

More and more, environmental compliance is an important requirement throughout the entire plant lifecycle from design to construction to operations and maintenance. Engineering design compliance involves satisfaction of requirements including exhaust pollution to atmosphere and toxic waste discharge to groundwater and surrounding soils. In some plants, special containment structures are required to mitigate effects from potential explosion and radiation hazards.

Safety and environmental compliances require systematic tracking of all safety and environment-related records in the form of data and documents. Good record keeping is mandatory so critical safety and environmental compliance information can be retrieved when needed. This particularly applies to long-term historical records. This is all part of information management.

## 4.4 Engineering Compliance

Engineering compliance satisfies engineering design criteria or design basis of the project that involves for not only the functional requirements of the plant but also for safety, environmental, local, and governmental regulations. Traditionally, such requirements are shown in various project documents as provided by the client. Such documents are usually in paper format, but even in electronic format it can be relatively difficult to interpret and to show compliance of the content. The preferred approach is to convert such documents into a design criteria database (DCD).

The DCD can be organized in a database program file (such as in Access) that contains all of the design requirements and references the appropriate document number, section and paragraph of the design criteria. The DCD is made up of three main parts:

1. The Criteria Requirements, which are identified by individual ID numbers

2. The Code Requirement, which is a listing of applicable codes for design
3. Implementation and Verification, which captures data indicating how, when, and who verified engineering compliance

The DCD provides a central source of design requirements by integrating design, safety, and environmental requirements into one source that can also be expanded into comparable databases for permitting, commissioning and operations. DCD is an object with a database attribute that is incorporated in an information data warehouse linked to design area, system, discipline, tag, and document objects (see Figure 4.1). The engineering disciplines are responsible for sorting and setting the requirements and then showing compliance. Required code documents can be a part of the document management system that is connected to the DCD.

**Figure 4.1 DCD Linkage to Information Blocks**

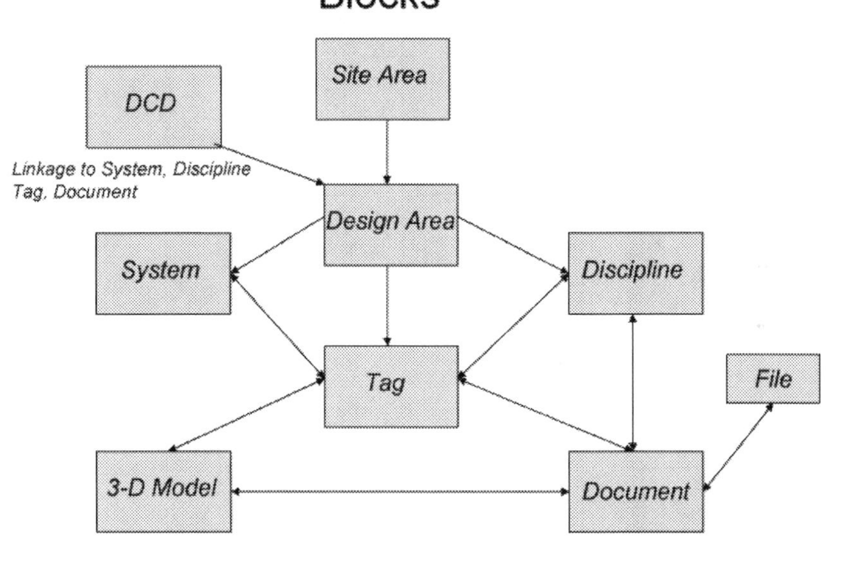


Through information management, DCD is a more systematic and accurate approach to show engineering compliance.

## 4.5 Configuration and Change Management

Changes are inevitable in the project execution of a process plant. The financial success of a project depends heavily on how well the changes can be managed. Tracking and recording such changes are another key part for project compliance. Currently, there are many paper-based, change-management systems that track design change requests, assessment, approval, and implementation. From an information management point of view, maintaining the distinction between the working environment and the information control environment is the first step to manage the what and when of change management.

In engineering design, changes can be managed at different times (i.e., at the end of preliminary design, at 40% detailed design completion, at base-lined release of technical information, when documents are released for procurement, when documents are released for construction, and so forth). All of these represent controlled information in a controlled environment and cannot be changed without some formal actions.

One change-management approach is the development of an intelligent design change request (DCR) database. Table 4.1 shows an example of a DCR form. DCR is an electronically generated template that shows the attribute fields. Once the form is filled in, the attribute information is loaded in the database. Related to DCR, there is also an impacted document change assessment form (IDCAF) as shown In Table 4.2. Again, this is an electronically generated template with attribute fields that become part of the DCR database once it is filled in. Other linkable forms can also be generated such as a DCR distribution list, DCR status reports, and so forth.

**TABLE 4.1 - DESIGN CHANGE REQUEST (DCR)**

Page ___ of ___

**DCR Number:** ______________ **Revision Number:** ______________

| | | | |
|---|---|---|---|
| Originator (Requested By): | ______________ | Date of Request: | ______________ |
| Subject: | ______________ | Discipline: | ______________ |
| Dwg/Spec/Ref: | ______________ | System/Area: | ______________ |
| Other Reference: | ______________ | | |

| | |
|---|---|
| Change Designation Design Development ☐ | EPC Initiated Change ☐<br>Client Directed Change ☐ |

**Change Description:**

**Reason for Proposed Change:**

**Change Justification:**

**Impact Assessment:**

| | | | | | |
|---|---|---|---|---|---|
| Schedule Impact | ☐ | ES&HQ Impact | ☐ | Cost | ☐ |
| Workhour Impact | ☐ | Operations/Commissioning | ☐ | Other | ☐ |

**Approval to Implement:**

| | | | |
|---|---|---|---|
| Lead Engineer: | ______________ | Configuration Contract Manager: | ______________ |
| Engineering Manager: | ______________ | Design Build Manager: | ______________ |
| Project Manager: | ______________ | DOE Concurrence: | ______________ |

Since DCR is an object with a database, it can be linked with design area, system, discipline, tag, and document blocks in the information data warehouse (see Figure 4.2). It can also be linked to a DCD if the client directs changes on the design criteria. The impacted documents are thus connected to the latest controlled versions, and once changes are implemented, the changed versions of tags and drawings are shown in the completed DCR. From the linkage, tags and 3-D models are also updated from the changes. DCR serves as a focal point for managing and recording engineering changes.

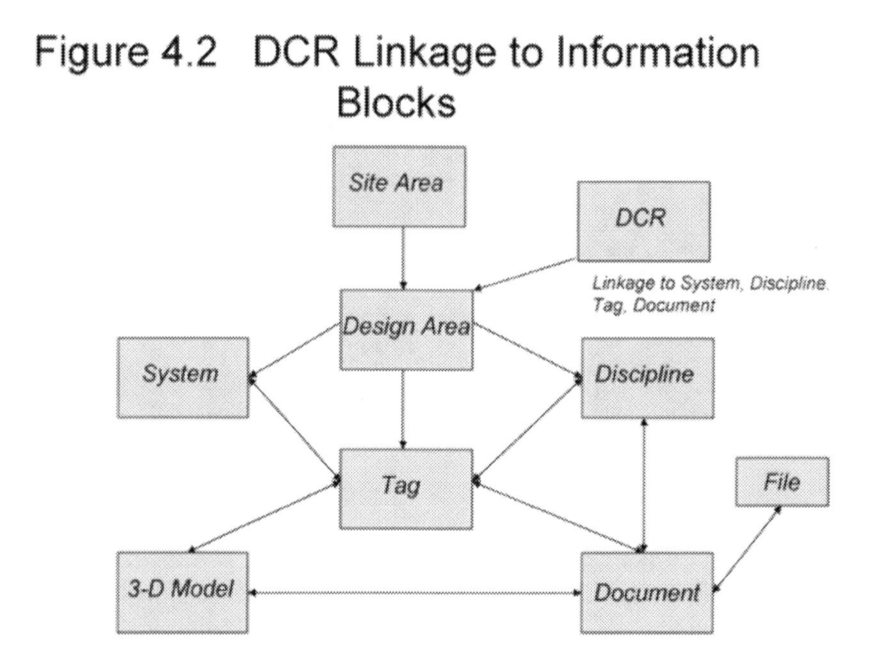


Figure 4.2 DCR Linkage to Information Blocks

Comparable change forms and the related databases can also be developed for change management of the procurement and construction processes.

**TABLE 4.2 - IMPACTED DOCUMENT CHANGE ASSESSMENT**

**DCR Number:** ____________

| | | | |
|---|---|---|---|
| Subject: | ____________ | Discipline | ____________ |
| System Area: | ____________ | | |

| | |
|---|---|
| Impacted Document Checklist (identify below if known) | Calculations ☐ |
| Basis of Design ☐ | Equipment List/MEL ☐ |
| Control Drawing ☐ | Interface Control Documents ☐ |
| PFDs ☐ P&IDs ☐ GAs ☐ Site Plans ☐ HVAC Flows ☐ One Lines ☐ | Others ☐ |

**Basis of Design (Section and Synopsis of Change Required):**

**Control Drawings:**

PFDs:

P&IDs:

GAs:

Site Plans:

HVAC Flows:

One Lines:

**Calculations:**

**Interface Control Documents:**

## 4.6 Procurement and Construction Verification

Project compliance for procurement starts with requisition and requests for quotation (RFQ) and requests for purchase (RFP). These documents contain the purchase requirements to suppliers, vendors, and fabricators for items related to performance, safety, environmental, and government regulations. Such requirements are generated from engineering documents based on design criteria and client specifications.

Before issuing a purchase order, suppliers go through a qualification process to ensure that they have the necessary QA/QC plans and procedures to satisfy the purchase requirements. Depending on the size and complexity of the purchased items, there are various supplier-data and document-checking steps by engineering along with inspections at the supplier company throughout the purchase cycles. This verifies that the purchased item satisfies the project requirements. The formal acceptance procedure accompanying final delivery of the product provides procurement compliance. All supplier information documents then are incorporated into the document management system, which is linked to the information data warehouse. Procurement is discussed in more detail in Chapter 6.

There are many steps throughout the construction process to verify that the plant is built as designed. The construction plan is the starting point to show how construction performance and safety can be satisfied, including construction worker qualification and training as part of a construction QA/QC plan. Extensive inspection and verification testing are required prior to formal construction acceptance tests to verify construction compliance. Construction job records must be included as a part of construction information management. Construction is discussed in more detail in Chapter 7.

## 4.7 Long-Term Information Storage

Project compliance relies on accurate project information records at every step of the plant lifecycle. A good project record system depends on what information needs to be recorded and how such information can be retrieved efficiently. With the modern technology of databases and low-cost information storage, there is now much more flexibility for developing a long-term information storage system.

The three basic information types—tags, documents, and 3-D models—when stored in an information data warehouse are, by definition, controlled information from the different stages of the plant lifecycle. Changes are marked and stored as different versions with version numbers and dates. Changed information is not deleted from the information data warehouse but is kept as long-term information storage. This is important because it provides traceability and tracks historical change records.

For example, a P&ID is released as controlled information to the information data warehouse as version A. Two weeks later, a version B with changes is released and two months later, a version C is developed with further changes. The three versions of this P&ID are kept as records, and any version can be recalled by specifying the version number and date. The same approach is applied to construction records. For instance, a pump is installed at a location at a given date and then, because of an engineering change, the pump is moved to another location at another date. The changes are also kept in the construction record, which is stored in the information data warehouse and can be recalled at any time with all pertinent information related to the move of the pump.

One can see that long-term storage and recall of information independent of the number of changes greatly enhances project compliance. Throughout the plant lifecycle, one may not only track each change in detail but even identify problems when they occur as well. Long-term information storage is one of the main advantageous features of information management in an information data warehouse. It is also a necessary requirement for project and plant compliance in a good information management system.

# CHAPTER 5
# PROJECT CONTROL

## 5.1 Overview

Project control is an important project function because the success of project performance is measured by two main project control jobs: managing schedule and managing costs. Like project compliance, project control does not generate information as in engineering and design but plans and manages project information on schedule and cost so the project can be executed smoothly throughout the plant lifecycle. There are many ways to do project control but the usual practice is first to assign a work breakdown structure (WBS) to the various work tasks so their performance in terms of schedule and cost can be tracked. The project control functions are applied throughout the project lifecycle from engineering and design, procurement, and construction through the mechanical completion of the project.

There are many good project schedule and cost control application programs. The success of project control depends on good planning and setup at the beginning of the project, close coordination with project disciplines, and an understanding of how information generated from the various project steps can be managed to fit into the project control programs.

Schedules are set up for various tasks and sub-tasks with practical and reasonable expectations of success to meet the project goals. Project costs estimated from the client requirements and preliminary engineering provide the bases for preparing budgets for the various tasks. Project control measures that project progress at various stages of the project lifecycle meets the planned project schedule and costs and identifies corrective actions for any deviations.

After planning the schedule and budget requirements, project control extracts performance information as the project progresses and manages the use of such information to ensure that requirements are being met. Reporting and record keeping are the main tasks.

## 5.2 Work Breakdown Structure (WBS)

The project WBS is a task and deliverable-oriented hierarchy of project elements that organizes and defines the total scope of the project. The WBS is a multi-level framework that organizes and displays elements representing work to be accomplished in logical relationships. Each descending level of the hierarchy represents an increasingly detailed definition of a project component.

At the beginning of the project, each project discipline plans its work functions in discrete steps. For engineering, these can be design tasks or subtasks, or it can be based on deliverables such as preliminary drawings or final reports. For procurement, the tasks can be the issuance of purchase orders or field acceptance of purchased items. For construction, the tasks can be the erection of a pipe way steel structure or completion of functional testing of certain equipment. Project control works together with project disciplines to assign WBS or sub-WBS numbers to all these tasks and functions. (See Table 5.1 for an example of WBS for construction).

For each discipline, output information from tasks and deliverable are connected to the discipline WBS. From an information management point of view, WBS is an object with a spreadsheet database that connects WBS numbers with attributes that define the tasks or deliverables. Thus, WBS is an object in the overall project information data warehouse (see Figure 5.1). WBS can then be connected to the discipline and system blocks of the information data warehouse in addition to tags, documents, and 3-D models.

While WBS numbers are planned together by project control and project disciplines to manage their works, they cannot, however, be too restrictive. For engineering changes, WBS can also be connected to DCS for direct assessment of changes to schedule and cost. In the field, there are many situations that may cause rearrangement of schedule, such as the late arrival of equipment and lack of proper construction personnel. With availability of proper information through the information data warehouse, there is more flexibility to react to changes at different WBS levels.

Table 5.1 WBS Construction

| WBS Code | WBS Name |
| --- | --- |
| C.2.3 | Construction |
| C2.3.1 | Construction Phase |
| C.2.3.1.00 | Construction |
| C.2.3.1.01 | Construction Project Management |
| C.2.3.1.02 | Technical Support |
| C.2.3.1.03 | Environmental, Safety, and Health |
| C.2.3.1.04 | Construction Procurement |
| C.2.3.1.05 | Construction Management |
| C.2.3.1.06 | Commissioning Management |
| C.2.3.1.07 | Management Plans/Docs |
| C.2.3.1.08 | Technical Plans/Documents |
| C.2.3.1.09 | Technology Development |
| C.2.3.1.10 | Process Building |
| C.2.3.1.10.01 | Process Cell Area |
| C.2.3.1.10.01.01 | Site Work |
| C.2.3.1.10.01.02 | Concrete |
| C.2.3.1.10.01.02.01 | Pre-Cast Concrete |
| C.2.3.1.10.01.02.02 | Mud Mat |
| C.2.3.1.10.01.02.03 | Cast In Place Concrete |
| C.2.3.1.10.01.02.04 | Equipment Pads |
| C.2.3.1.10.01.03 | Structural Steel |
| C.2.3.1.10.01.03.01 | Erect Deck Steel |
| C.2.3.1.10.01.03.02 | Install Cell Liner |
| C.2.3.1.10.01.03.03 | Erect Secondary Steel |
| C.2.3.1.10.01.04 | Doors, Windows, and Finishes |
| C.2.3.1.10.01.04.01 | Dry Wall |
| C.2.3.1.10.01.04.02 | Painting |
| C.2.3.1.10.01.04.03 | Roofing |
| C.2.3.1.10.01.05 | Mechanical |
| C.2.3.1.10.01.05.01 | HVAC |
| C.2.3.1.10.01.05.02 | Plumbing |
| C.2.3.1.10.01.05.03 | Piping |
| C.2.3.1.10.01.05.04 | Tanks |
| C.2.3.1.10.01.05.05 | Equipment |
| C.2.3.1.10.01.06 | Specials |
| C.2.3.1.10.01.07 | Electrical, Controls, and Communication |
| C.2.3.1.10.02 | Cold Chemical Area |
| C.2.3.1.10.03 | Facility Support Area |
| C.2.3.1.11 | Administration Building |
| C.2.3.1.12 | Diesel Generator Building |
| C.2.3.1.13 | Compressor Building |
| C.2.3.1.14 | Stacks |

| | |
|---|---|
| C.2.3.1.15 | Yard |
| C.2.3.1.16 | General Conditions |
| C.2.3.1.17 | Construction Testing |
| C.2.3.1.18 | Alpha Finishing Facility |
| C.2.3.1.19 | Other Direct Costs |
| C.2.3.1.20 | Management Reserve |
| C.2.3.2 | Construction Design Support |
| C.2.3.2.02 | Construction Project Management Support |
| C.2.3.2.03 | Environment, Safety, and Health |
| C.2.3.2.04 | Procurement |
| C.2.3.2.10 | Safety Analysis |
| C.2.3.2.12 | Construction Design Support |
| C.2.3.2.13 | As-Builts |
| C.2.3.2.14 | Process Engineering Support |
| C.2.3.2.90 | Other Direct Costs |

Figure 5.1 WBS Linkage to Information Blocks

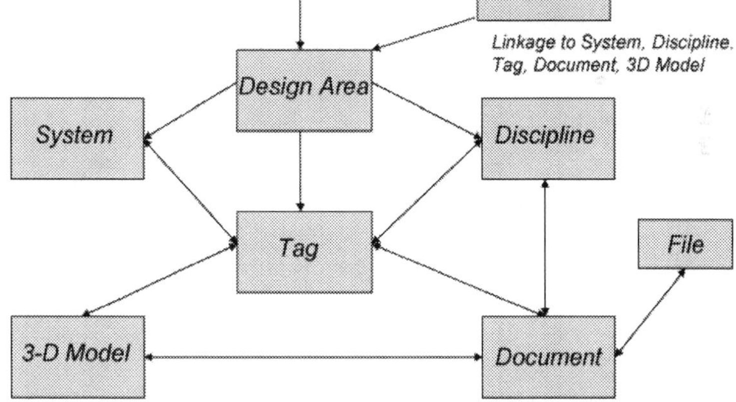


## 5.3 Schedule and Budget

Project control and project disciplines work and plan together an overall schedule to complete project tasks. WBS numbers are linked to the schedule numbers of the required schedule levels as defined in a good schedule program. The schedule requirements for the discipline and the relationships to other disciplines are also defined.

As the project progresses, information is generated and loaded into the information data warehouse as shown in the three basic information types. For example, different stages of task completion for 3-D models for each discipline are linked to the discipline's WBS and then connected to the schedule program. Similarly, progress in deliverables as drawings or purchase orders are recorded through the WBS as progress in the schedule program.

In a financial program, budgets are usually prepared for various project tasks and deliverables in terms of labor and non-labor (purchased item) costs. For each discipline, man-hour estimates are made in the beginning of the project; labor costs are then estimated per labor rates and budgeted for the discipline's tasks accordingly. A WBS is then assigned to the tasks and cost budgets are linked to it through cost numbers as defined in the financial programs. The same approach is used for non-labor cost items. Cost budgets are prepared for purchased (usually tagged) items. These cost budgets are related to WBS numbers and linked to cost account numbers in a financial program.

Schedule and budget are then managed through established schedule and financial programs. However, progress in different areas of the project generates information identified by tag numbers, deliverable documents, and 3-D models that are fed into these programs. The key consideration here is how to link and track schedule and cost numbers to WBS information. Project performance depends on the successful reporting and recording of such information.

## 5.4 Cost Control and Report

Once project performance information on schedule and cost has been gathered, project control serves to compare that information against the established schedule plan and budget. This requires establishing the bases for assessment at the beginning of the project. From a project schedule point of view, for example, what '50 percent design completion' means must be defined. Forty percent completion could be defined as the first release of P&ID for review. On piping 3-D design, 70 percent completion may be at piping isometrics extraction. It is important to have unified and acceptable methods for such progress assessment.

Depending on different industries, there are different ways for project control to assess and report project progress. For example, on some government projects, the earned value management system (EVMS) is used. Again, the key point is to have accurate information on the status of schedule and the actual expenditure on man-hours and costs so that the results can be recorded and reported.

Any project changes—whether at the design stage or at the construction stage—have impacts on both schedule and cost. Hence the DCR form for engineering shows the requirement for line-items impact assessments on schedule and cost for any changes. With linkage of WBS to DCR, these changes can be more easily coordinated.

Information management integrates project control as a part of the total project production information environment. Project control links project schedule and cost information with the three basic information types in the information data warehouse. This ensures that project control uses the correct information with guaranteed integrity throughout the project lifecycle.

# CHAPTER 6
# PROCUREMENT

## 6.1 Overview

Procurement purchases bulk material and all tagged items, such as pumps, instruments, and equipment, for the project. Procurement also issues subcontracts both for engineered and fabricated items—such as steel structure components, piping spools, special equipment, and so forth—and also for subcontracted consulting services, constructors, and so forth. Procurement performs three basic activities that are important to the project lifecycle because project cost and schedule depend on the successful performance of these activities.

First, procurement maintains vendor relationships. To prepare for purchase, procurement keeps knowledgeable on current market price information for commodities and products and maintains a listing of qualified vendors on their products and qualifications. Second, procurement interfaces with engineering to gather the information necessary to prepare for requisition and the issuance of purchase orders. Third, procurement identifies what will be required for successful execution and fulfillment of purchase orders.

The first activity is mostly in the procurement's working environment, which means the activities are performed within the procurement discipline with limited interface with other disciplines. For the second activity, project procurement has a strong information linkage with both engineering and project control with respect to cost, budget and schedule requirements. Procurement is responsible for the successful completion of purchased items that meet specification, cost, and schedule requirements. Procurement then interfaces with the construction for delivery of the purchased items to the field along with completion of MRR (materials receive report). Vendor documents as part of procurement deliverables are also required for construction and comprise a part of hand-over at project completion needed for the operations and maintenance stage of the plant's lifecycle.

Since procurement activities involve extensive information management, we want to structure these activities as objects with databases so that procurement information can be linked efficiently with other project objects from engineering and project control through the information data warehouse.

## 6.2 Procurement Main Information Flow and Objects

From an information management point of view, Figure 6.1 shows one approach for setting procurement activities as objects and the relationship of information flow between these objects. The four main information blocks are requisition (tracked by requisition number), purchase order (tracked by purchase order number), vendor (tracked by vendor number), and all MRRs ( tracked by MRR number).

**Figure 6.1 Procurement Main Objects & Information Flow**

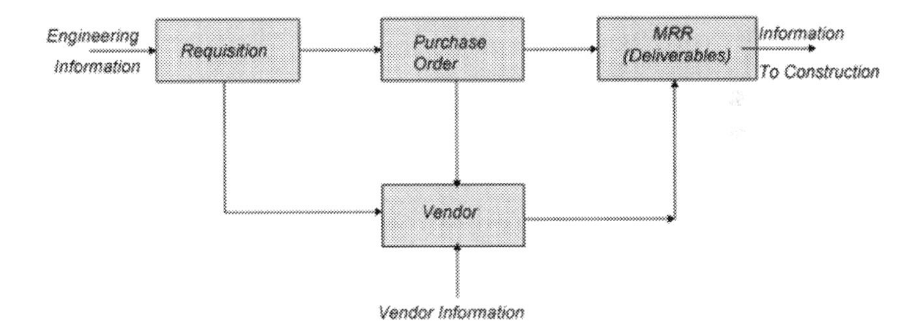


The requisition gathers and assembles engineering information into a set of requirements for procurement. Ideally, the document would be generated from a template that consists of attributes linked to engineering information extracted from the information data warehouse. Requisition may also be generated for sub-contracts. In this case, the format and contents may be in the form of a scope of work. The first connection of requisition to vendor is through RFQ (Request for Quotation) or RFP (Request of Proposal). Based on the requisition, an RFQ obtains product specification, performance, and cost from the vendor with no commitment to purchase. On the other hand,

RFP is a formal request for the vendor to bid on a requisition given the proposed product, cost, and schedule. The vendor's bid has a validation time period for the proposal. RFQ and RFP are usually sent to several qualified vendors.

The bids from vendors are then evaluated for both technical and financial adequacy. Discussions and meetings may be necessary to clarify questions and through detailed evaluations, a vendor is selected for the purchase based on the best technical and financial proposal.

A purchase order is then issued to the selected vendor for the purchase. A purchase order is a document consisting of several parts. In addition to terms and conditions for the purchase, it contains technical data, shipping, and schedule requirements usually already part of the requisition. Like the requisition, a purchase order ideally could be generated from an intelligent template that links and places attribute information from the requisition in it. For sub-contacts, a purchase contract with a scope of work may be used instead of standard purchase orders.

There are several activities between the purchase order and the delivery of purchased items from the vendor. There are expediting and traffic functions. Inspection and shop testing may be part of the purchase order requirements. Once the purchased items are delivered and accepted in the field, an MRR is generated. This document is the information linkage between procurement, the vendor, and the construction. It also verifies that the deliverables, including the physical items and vendor data, satisfy the requirements as specified in the purchase order.

A vendor may be set up in the information data warehouse as an object representing information linkage from procurement to the vendor. Procurement performs vendor management, which includes vendor qualification, maintaining a qualified vendor listing, maintaining a current pricing and availability listing, and other tasks. Procurement also handles information flow from requisition to vendor through RFQs or RFPs, bidding from vendors, purchase order issuance to the selected vendor, and monitoring the successful execution of the purchase order from the vendor. Once the vendor delivers the purchased item, an MRR is completed confirming that the delivered item meets all requirements in the purchase order.

There are several good procurement computer programs that generate requisition, purchase order, and MRR documentation. Some programs also contain modules for vendor and materials information management that are used mostly in construction. These four main procurement objects are included as part of the project information data warehouse that provides linkage to other objects in the warehouse. The following paragraphs show more details on these information linkages.

## 6.3 Requisition

Typically, a requisition may consist of many sections. There are two sections: the deliverable items list (DIL), the technical document and attachment (TDA) that are linked to engineering objects. There is a section on supplier data requirements (SDR). Other sections typically concern quality level requirements, source inspection requirements, a materials control list, shipping, and other details.

The DIL defines the main purchase requirements reflecting the attributes of the purchase item—including item number, tag number, description, quantity, unit price, promised ship date, extended amount, and the total amount. The TDA includes required drawings, specifications, datasheets, listings, tag data, 3-D models, and any other required items that have been extracted from the information data warehouse as controlled, single-source, accurate, and current information.

Figure 6.2 shows the requisition and the related requisition list information that is linked to other project objects. It connects to basic engineering information through tags, documents, and 3-D models. It relates to the system block to show the project requirements through job number and sub-job number and links to the discipline block to identify the requisition originator and the approval routing requirements. Through linkage to WBS, it connects to schedule and cost requirements such as required date at site, budget estimate, and so forth. The requisition is the basis of information for a vendor to bid on the purchased items. The requisition may go through RFQ, RFP, bid conference, or some other method appropriate to the project. After bid evaluation, requisition provides the basic information for the issuance of a purchase order. Requisition may also contain a section on scope of work if issuance of a purchase order for a subcontract is required.

# Figure 6.2 Requisition Information Relationship

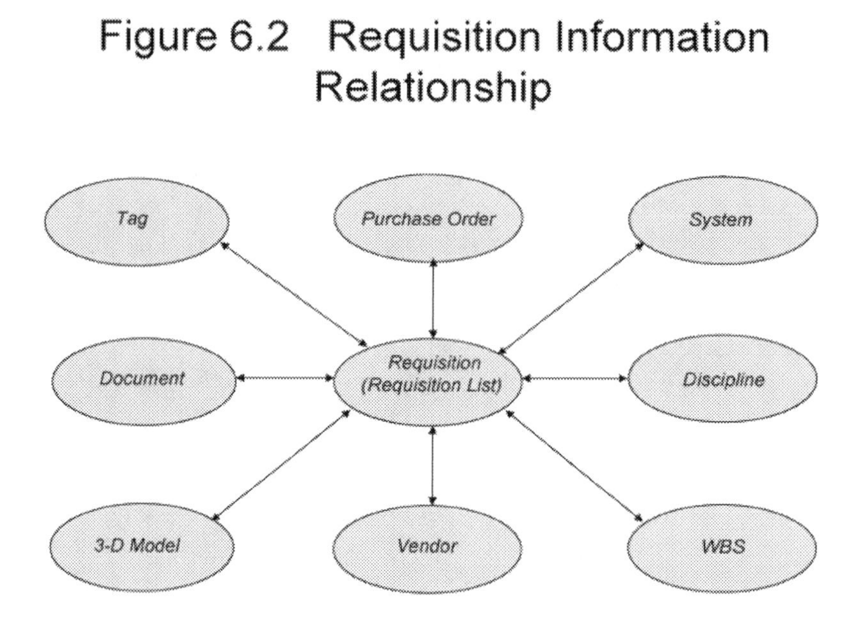


Once a requisition is issued, it may be impacted by engineering changes through DCR. However, because of linkage to engineering objects that show the engineering changes a requisition can always pick the correct and current engineering information.

## 6.4 Purchase Order

Following bid evaluation, a successful bidder is selected, and a purchase order is issued to the successful vendor. A purchase order follows the general format of a requisition with the addition of sections on contractual, financial, shipping and other terms and conditions. The DIL, TDA, and SDR all can essentially be copied over as part of the purchase order. Additionally, the purchase order can be an intelligent document generated from a template via associated attribute database information.

For successful execution of a purchase order, procurement performs activities such as traffic, expediting, vendor shop inspection, shop acceptance testing, and so forth. Information on these activities is documented, but once the purchase order is completed for these items, the documents are kept as records and have limited usage to other life-cycle functions.

Figure 6.3 shows the purchase order information relationship with other objects in the information data warehouse. The discipline block tracks purchase orders by purchase order number, while the system block tracks purchase orders by job number and sub-job number. The main purchase order linkage is to requisition. Through the requisition, it connects to engineering objects of tag and document. The purchase order performance on cost and schedule is managed by project control and is linked through WBS to schedule and financial programs. Procurement has direct contact with vendors through purchase orders because they contain vendor information such as vendor address, contact person, and so forth. Procurement also manages vendor relationships on successful execution of purchase order that all required deliverables were on schedule and within budget. At the completion of a purchase, purchase order information is connected to MRR and provides the basis for verifying that the delivered physical items and the required documents satisfy the purchase order requirements. Purchase order revision follows requisition revision and has the same formal steps as DCR for engineering changes as explained in the previous section.

Figure 6.3 Purchase Order Information Relationship

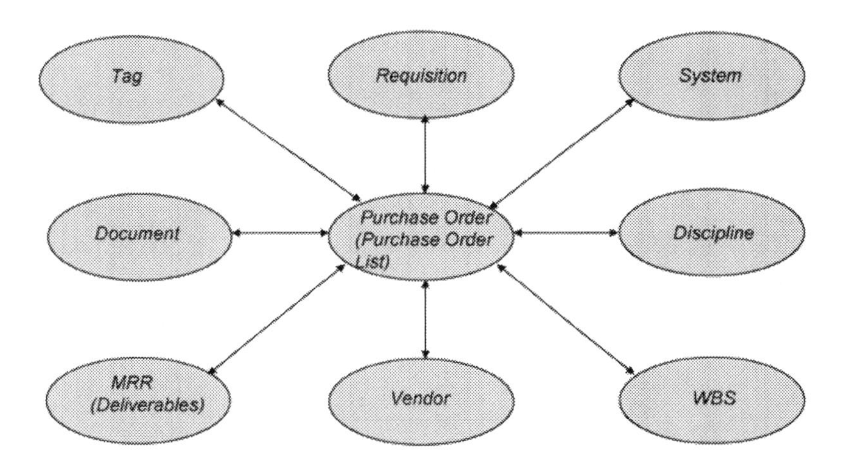


Purchase order for a subcontract is different depending on the types of subcontract. If a subcontract is for fabrication of engineering components such as piping isometrics, steel support structure, and so forth, then engineering has more direct contact with the fabricator. In some cases, engineering manages this type of subcontract. If a subcontract is for part of construction, then the overall construction management has direct control of the contract. The information from deliverables and documents, however, is linked to other objects in a similar manner as purchased items.

Purchase order can be an intelligent document that it in an object with database and contains main project information that links engineering to procurement to construction. Purchase order revision follows requisition revision and it has the same formal steps as DCR for engineering as explained in the previous section.

## 6.5 Vendor

We use the general term vendor here to include supplier, fabricator, or subcontractor. Procurement has a strong interaction with vendors. Before issuance of a purchase order, procurement activities include vendor contact, qualification, bid solicitation (RFP/RFQ), and bid evaluation. Usually, a vendor number is assigned to track the vendor information like vendor capabilities, contact point, and all related vendor functions. Computer programs such as the online vendor management (OVM) system are available to manage vendor information. The pre-purchase order activities, however, are basically in the procurement team's working environment. Once the purchase order is completed, information is kept as long-term records but has limited use in the overall lifecycle applications.

Figure 6.4 shows the vendor and its relationship to other objects as an object in the information data warehouse. It shows how vendor links with requisition and purchase order, both of which define procurement requirements containing the necessary tags, documents, and 3-D model engineering information. This especially concerns sub-contracted vendors because they have a more direct interface with engineering information. For execution of purchase orders, the vendor item provides schedule and cost information linked to WBS and hence the project schedule and financial programs as well.

# Figure 6.4 Vendor Information Relationship

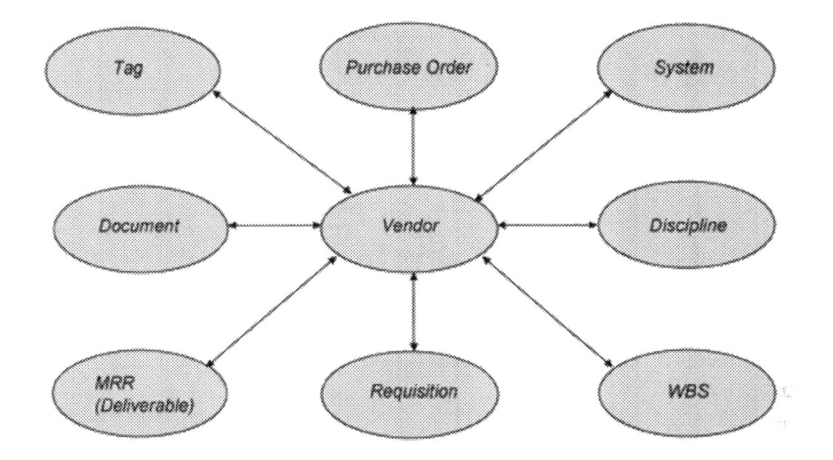


On the delivery of purchased items, the vendor provides information that populates an MRR. One major item is a vendor data submittal, which includes drawings, installation details, operating manuals, maintenance procedures, spare parts requirement, and so forth. While vendor data requirements are defined in requisition and purchase order, the usage itself is in construction, mechanical completion, start-up, and operations and maintenance.

## 6.6 Material Receipt Record (MRR)

When the purchased item and the related vendor data are delivered and accepted in the field, a MRR is generated. This may include other reports such as OS&D (overage, shortage, & damage) and LDD (lost, destroyed, damaged) that are further related to MRR. MRR is considered as an object because it is the end of the procurement function and is the functional linkage between procurement and construction.

MRR is an intelligent form that contains attribute information linked to other objects in the information data warehouse (see Figure 6.5). The form contains the project number linked to the system and discipline blocks along with receiving information. MRR is tracked through an MRR number and the issue date. It is linked to requisition and purchase order (PO) through requisition number and PO number. Whether the purchased item is bulk material, a fabricated item, or a tagged item, MRR identifies the purchase order item, its tag number, its description, and the quantity ordered. MRR links with vendor to show vendor number and name, received date, received quantity, receiving status, and storage location. The delivery date in MRR and the subsequent vendor invoice are linked to WBS and project control for schedule and financial information.

The list of MRR provides the basic information as inputs for inventory control in the construction warehouse.

Figure 6.5 MRR Information Relationship

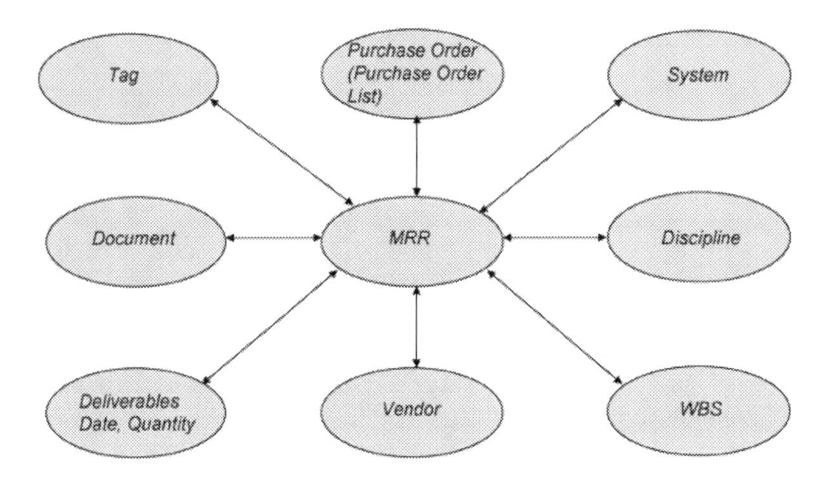

# CHAPTER 7
# CONSTRUCTION

## 7.1 Overview

Construction is a main step in the plant lifecycle as it may comprise more than 40 percent of the project cost. It interfaces with engineering and procurement in the front and mechanical completion and plant operations in the back. The successful execution of construction determines the overall financial and schedule success of the project and establishes the subsequent efficient operations and maintenance of the plant. Figure 7.1 shows the main construction activities that start with construction planning. Field receiving, inventory management in the warehouse, and issuing of purchased materials and items are construction activities that interface with procurement. The actual construction work is represented by construction and installation activities. Construction inspection and acceptance tests are conducted to verify that the plant is built as designed and that the purchased equipment and items perform as specified. There is also the major activity of gathering and assembling construction information and documents as assembled in a construction information listing after mechanical completion to prepare for the final acceptance of the constructed plant and hand-over to the owner or operator to begin plant operations and maintenance.

Construction activities generate a large quantity of documents and information. But, most of these are for records and references for future use when needed. A good document management system is required for record keeping. However, there are several types of construction-related information, which are organized as objects with databases linked to other objects in the information data warehouse. The following paragraphs explain in more detail the information relationship of construction activities.

Figure 7.1 Construction Activities & Information

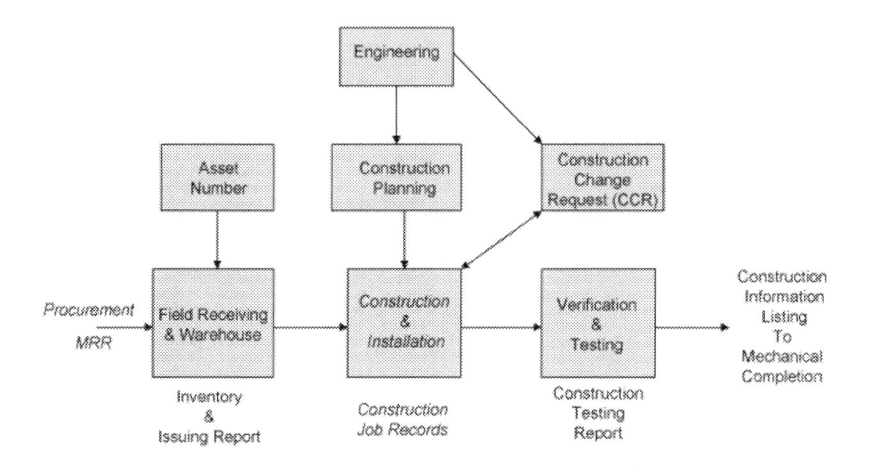


7.2 Construction Planning

Construction planning should start as early as possible, preferably with the onset of engineering and design activities. Construction planning interfaces with engineering through engineering documents, tags, and especially with 3-D models (see Figure 7.2). The 3-D models provide a powerful tool for construction planning with respect to construction sequences, space allocation requirements, construction equipment requirements, and so forth.

Figure 7.2 Construction Planning

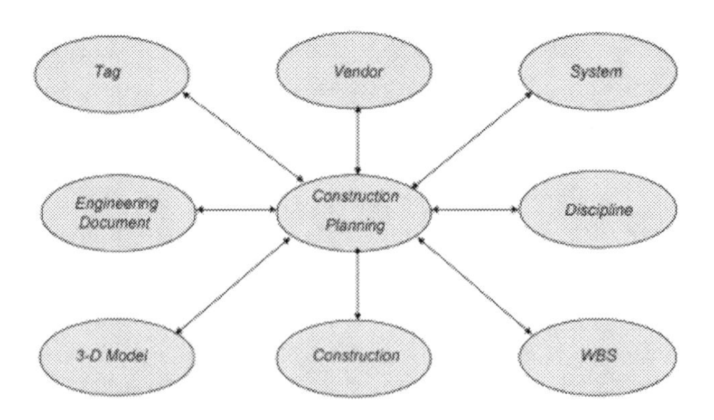


A constructability study is also a part of construction planning. For example, for a process plant with high bays, availability of heavy capacity and high level lifting equipment may impact on the decision to use pre-stressed concrete roof beams versus cast-in-place concrete beams. Equipment weight and size may be restricted on certain plot plans and piping layouts because of available space and construction sequence considerations.

The construction team works with project control on developing construction WBS that controls construction schedule and costs for all construction-related activities. Utilizing project job numbers and sub-job numbers, the construction team plans the work for different systems and disciplines at different parts of the plant.

The team may also interface with suppliers and fabricators through procurement to obtain advance information on equipment and items for planning. It coordinates with sub-contractors if so required to obtain information from engineering for planning.

It also produces planning documents that are used directly in construction activities. While such documents contain accurate and current engineering information extracted from the information data warehouse, they are not intelligent objects with databases. These documents are managed by document control under a construction-planning heading.

## 7.3 Field Receiving, Warehouse Inventory, and Issuing

All the purchased materials, equipment, instruments, parts, and fabricated items delivered to the construction site are received by construction and properly documented. The MRR is key here, as its organization as an object provides the information interface between procurement and construction.

At this point, a new object—an asset number—is developed and assigned that links MRR, field receiving, and tag number (see Figure 7.3). In specific industries, asset number may be referred to as plant tag number or government property number. The meaning as explained below is nevertheless the same.

Where tag number is an engineering term that defines the functional properties of some equipment, device, or item at its given location on a plant design as represented in drawings and 3-D models, an asset number represents a physical item with the functional properties of its corresponding tag and more. It also contains unique information that identifies the physical unit such as manufacturer name, unit model number, serial number, and so forth.

Asset number is important for construction installation and for downstream operations and maintenance. For example, a pump has a tag number to identify its properties and its location in an engineering design. Once that pump is purchased and installed at a location by construction, it has an asset number to identify that it is a unique physical item. The pump's asset number then can be used to link the maintenance history of the item. Also, the pump at that location can be replaced with an identical spare pump or a spare pump from a different manufacturer with different asset number, but the tag number will remain the same.

## Figure 7.3 Field Receiving and Warehouse

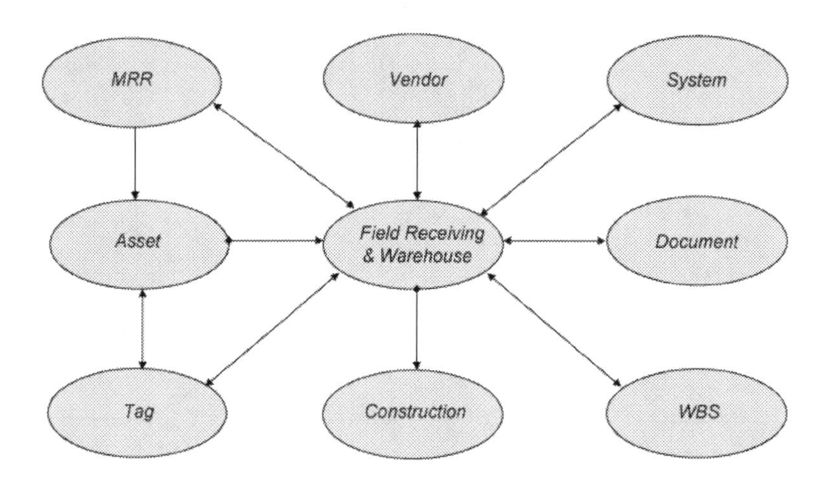


In construction, a tag number may be used to identify fabricated items such as piping isometrics or piece mark number for steel structural components. An asset number may not be required in such a case. Asset number is an object with attributes linked directly with data information from a tag number. Additional attributes are added to the asset object to show its unique physical identity.

Purchased materials and items stored in the warehouse are managed and issued for construction as needed. Documents that record these activities are kept in the document management system available for viewing in the subsequent project functions. However, inventory records of spare asset items and all vendor data are assembled and then turned over as an information package to plant operations and maintenance.

Field receiving interfaces with vendors through procurement to resolve any delivery issues such as shortage, damaged items, and so forth. The construction WBS includes field receiving, warehouse, and so forth, so the schedule and costs of these activities can be tracked by project control.

### 7.4 Construction and Installation

The main construction and installation activities involve resources management, construction personnel assignment, rigging design, space allocation, construction equipment requirements, construction sequencing, progress reporting, and other required project activities. Computer programs are available to manage these functions. From an information management perspective, Figure 7.4 shows the information linkage of construction and installation to other project objects.

Figure 7.4 Construction and Installation

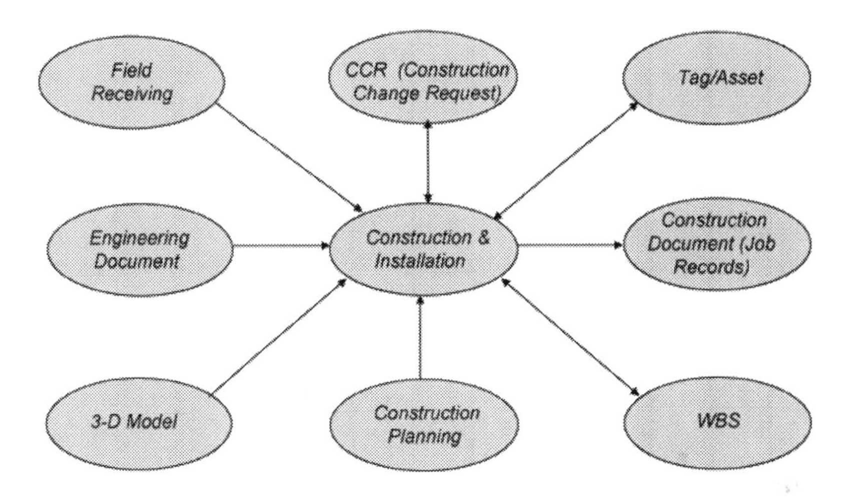


Construction planning links directly to the execution of construction and installation. Engineering information as used in planning activities is referenced and used in construction. Three-dimensional models are particularly useful for visualization and space allocation for installation. The models are also used for construction progress tracking and reporting. Vendor supplied equipment, instruments, and components as asset items in relation to tag number are installed according to vendor data on installation requirements.

The construction team's schedule and costs are fed to project control through the team's WBS. While construction reacts to engineering and procurement changes, there are also construction changes initiated in the field. In this case, information feeds back to procurement, project control, and engineering for proper coordination and documentation. Construction change requests (CCR) follow the same approach as DCR with information linkages to engineering and procurement objects and to project control objects for schedule and cost impact.

Construction and installation produce large quantities of information and documents. The construction job record (CJR) is a construction journal that records daily, weekly, and monthly construction-related activities. It is not treated as an object but is well organized and indexed in a good document management system so that it serves as a central point for linkage to other construction documents. Overall, the two main objects in the construction and installation function are the asset number and CCR.

In some projects, construction sub-contractors are used to do certain parts of construction work. Construction first coordinates sub-contractors with engineering, field receiving, and project control and then manages the successful performance of construction works by the sub-contractors.

## 7.5 Verification and Testing

Construction requires inspection and testing to verify that the plant is constructed as designed and that systems, equipment, instruments, and items perform properly to satisfy the functional and safety requirements of the plant. Construction inspection is a continuous process that is done as each construction component and subcomponent is completed. This is also a part of construction QA/QC procedures for project compliance. The results of inspection are thoroughly documented and kept as part of the construction records.

The construction acceptance test (CAT) is a formal step in verification and acceptance of constructed components in concrete, steel structure, piping, electrical, instrumentation, and control. It is also done at the system level to show that equipment, instruments, and components in a system perform as a unit to satisfy the functional requirements of the project. CAT is organized as an object with a database linked to engineering information, vendor data, and asset information for producing testing requirements, a testing plan, and procedure documents (see Figure 7.5). CAT is a summary verification that construction activities satisfy project requirements. The test results are documented and form the planning basis for the systems operational test (SOT) during the next lifecycle stage. Verification and testing interfaces with project control through WBS to track schedule and costs for these activities.

Figure 7.5 Verification & Testing

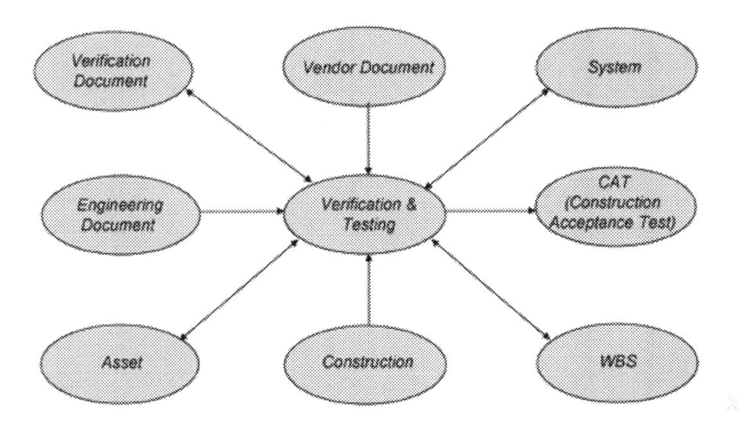


Welding is a key activity in the construction of steel-structure and pressurized-piping components. For safety consideration, welding verification has developed a more formalized procedure. Welds are tracked through weld tags and weld maps, while welding procedures along with weld X-ray and ultra-sound testing results are fully documented and kept as records.

## 7.6 Construction Information Listing (CIL)

A construction information listing (CIL) summarizes all information and documents covered by construction activities. Figure 7.6 shows the four new construction related objects—MRR, asset number, CAT, and CCR—that are linked to other objects in the information data warehouse. The CIL also identifies construction-related major document types, including construction plans, vendor data (documents), inventory and issuing reports, construction job records, and verification records (QA/QC). The construction-related documents are stored in a good document management system that can be effectively tracked, reviewed, retrieved, and published.

## Figure 7.6 Construction Objects & Document Types

| Linkage to Existing Objects | New Construction Objects |
|---|---|
| System | Asset Number |
| Discipline | MRR (Material Receiving Records) |
| Tag | CAT (Construction Acceptance Test) |
| Document | CCR (Construction Change Request) |
| 3-D Model | |
| WBS | **New Construction Document Types** |
| Purchase Order | Vendor Data (document) |
| Vendor | Inventory & Issuing Report |
| | Construction Job Records |
| | Verification Records (QA/QC) |
| | Construction Information Listing |

Construction also identifies and marks all as-built information transferred to engineering for updating documents, databases, and 3-D models in the information data warehouse. This updated information is in the controlled environment and may therefore be used with confidence in downstream activities.

At the end of construction, key information as content in the CIL is transferred for use in the next stage of the plant lifecycle—mechanical completion and hand-over.

# CHAPTER 8
# MECHANICAL COMPLETION AND COMMISSIONING

## 8.1 Overview

This chapter describes the activities between the completion of construction and the commissioning and hand-over to the owner or operator of the plant for start-up of the plant. The activities follow either by construction areas and sub-areas or by systems as determined by process sequences. Owners or operators sometimes participate as part of these activities.

Figure 8.1 shows the overall steps in commissioning plan that leads to plant hand-over and start-up. Systemization is the first step. This activity links all process and engineering information by systems to the physical components of the system, including process equipment, instrumentation and control, and interconnecting pipes, conduits, and wires. Physical facility support systems include components such as facility equipment, electrical supply, HVAC, structural support, and building.

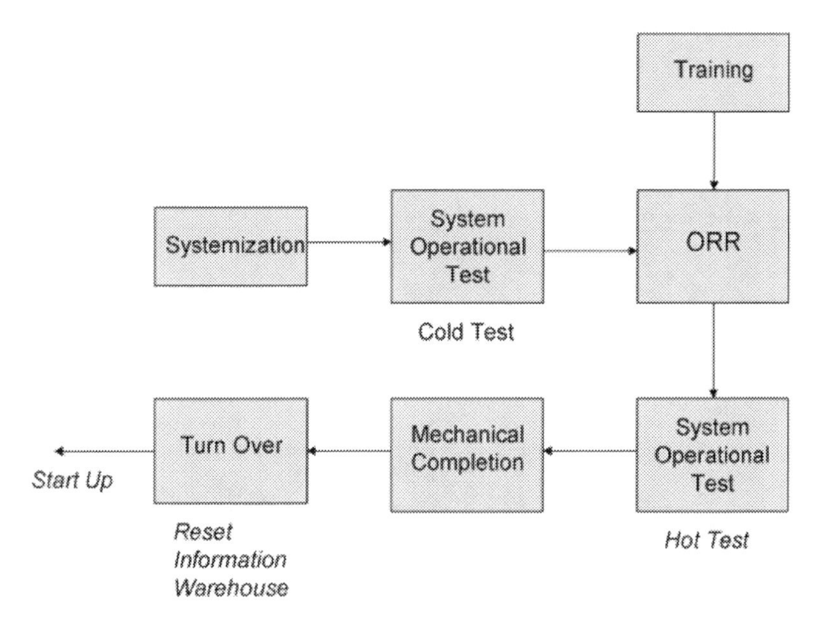


After systemization, the next step is to plan and perform a system operational test (SOT) on the defined process systems. The tests may first be a series of 'cold' tests at this level, which means that they may not be at full operational levels for heat and power inputs to cause all required processes or chemical reactions to occur. Results of previous CAT along with systemization comprise the basic inputs for SOT. These tests verify that the system components themselves and their linkages to other components work and satisfy the basic system safety and operational requirements.

In some process facilities where safety and hazards are prominent considerations (e.g., in a nuclear facility or at an explosive process facility), an operational readiness review (ORR) may be a major step between 'cold' and 'hot' system operational tests. Different process industries may use different terminology, but ORR indicates a critical review of all process and engineering information along with results from cold SOT tests with a special emphasis on safety and hazards to determine if systems are ready for start-up operations. Information on training and qualification of plant operators is one input to ORR as it demonstrates sufficient qualified operators to start-up and run the plant safely.

Once the system passes ORR, it is ready for a 'hot' system operational test. The test may go through several incremental steps to reach the full system design power level. The results of this test (or these tests) are given to the relevant interested agencies for issuance of an operating permit or a license for plant start-up.

These three elements—systematization, system operational test, and operational readiness review—constitute commissioning objects with databases that are linked in the information data warehouse (see Figure 8.6).

The next step of commissioning is the hand-over of the plant. This has traditionally been a major interface between the EPC and the owner or operator. The proper planning of information hand-over should begin as early as possible, ideally at the engineering and design stage. Preparation of an efficient information management system for hand-over can have potential saving of up to 5 percent on the total capital cost of the plant. Proper hand-over of the plant as-built information can also mean considerable cost saving through subsequent efficient plant operations and maintenance.

## 8.2 Systemization

Systemization is a planning and preparation step at the system level to integrate process and engineering information with physical process equipment and components information. As shown in Figure 8.2, systemization is first linked to systems as defined in the engineering and design phase for the process-related systems and facility-related systems. Process and engineering information—represented by the three basic information elements of tag, document, and 3-D model and stored in the information data warehouse—are used to define the system relationships and performances.

# Figure 8.2 Systemization

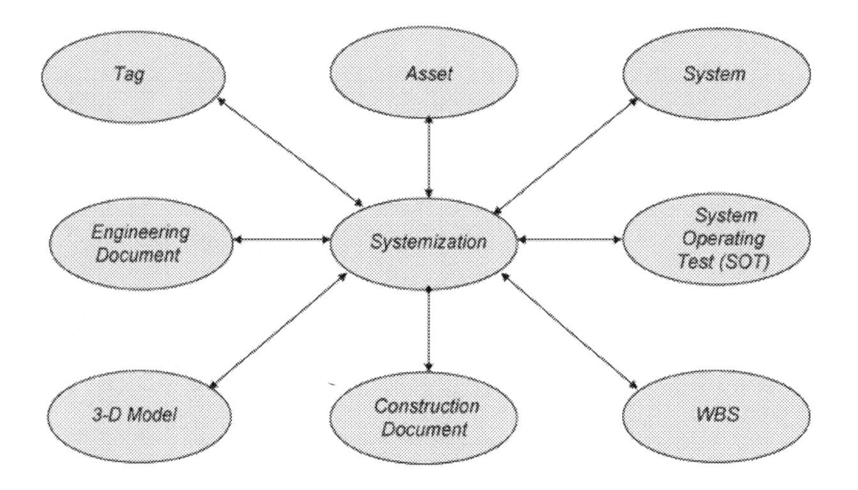


For process systems, the planning starts at the process flow diagram (PFD) and P&ID levels and expands into line summaries and control systems. All associated components in the system such as process equipment, control valves and instruments are identified by their tags and related databases. Components shown as physical items in construction and installation are linked from tags to asset numbers and their associated databases.

The performance requirements in terms of process parameters for temperature, pressure, flow rate, product specification, and so forth are identified for normal operations, abnormal events, and emergency conditions. Any special items for safety and hazard requirements are also listed. Systemization of process systems includes sub-systems, systems, and integration of production systems for the complete process plant.

Facility systems such as facility electrical, HVAC, fire protection, and so forth follow the same approach—retrieving controlled engineering information from the plant data warehouse and linking it with asset-related physical items. For facility systems, systemization may not be as rigorous as for the process systems, but it is just as important for the performance and safety of the complete plant.

The result of systemization is a set of intelligent documents and their related databases used as inputs for the preparation of System Operational Tests (SOT).Again, WBS is linked via WBS to project control schedule and cost activities.

## 8.3 System Operational Tests (SOT)

All pertinent information gathered by systemization is fed into planning and preparation for SOT. At this point, process, facility equipment and components performance has been checked out and accepted in CAT, which provides an input to SOT (see Figure 8.3) along with engineering and design information from tags and documents.

**Figure 8.3 System Operational Test (SOT), Cold and Hot**

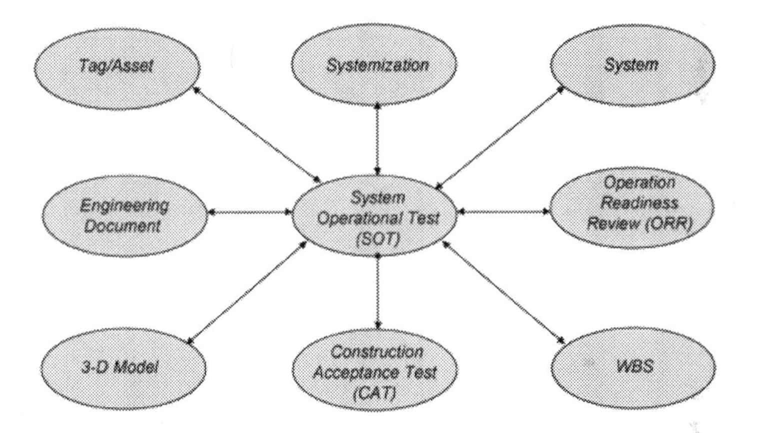


SOT follows a step-by-step approach from sub-system to system to the complete process facility. For 'cold' SOT, operating conditions gradually increase from low levels to higher levels with thoroughly documented and evaluated test results before the next level of testing starts. For some process facilities, tests are carried out to cover the full spectrum of operating conditions from normal to abnormal and emergency conditions.

As stated earlier, for some critical process facilities, ORR provides an intermediate step between cold and hot tests and the hot test of the full operating conditions. This may be necessary for safety or hazard reasons prior to a full performance level test. In this case, cold test results are intended ultimately to serve as inputs for ORR. With ORR results and an assurance of safe performance, testing then advances to hot tests and finally full operating conditions for the process facility.

At every step, the test results are evaluated and thoroughly documented in preparation for hand-over and start-up. Results are also summarized in the appropriate format as a part of submittal process to government or regulatory agencies that will issue licenses or permits for facility operations.

## 8.4 Operation Readiness Review (ORR)

As the term implies, ORR constitutes a formal review of the project to ensure that the constructed process plant is ready for start-up and operations. While considerable emphasis is placed on review and acceptance of the operational safety of the plant, ORR also is a summary review to ensure that the plant is designed and engineered to satisfy the performance and safety requirements of the owner or operator, that the equipment, components, and materials have been purchased as specified, and that the plant is constructed as engineered.

The success of ORR depends on the orderly and efficient availability and traceability of information. So far, the use of an information data warehouse has emphasized its effective means for managing plant project lifecycle information via linked objects with databases. For engineering design, we have the basic information objects of tags, documents, and 3-D models. For procurements, we have the objects of requisitions, purchase orders, vendor information, and MRR. We have the objects of asset numbers and CCR (for construction), along with other non-linked, construction objects (inventory and issuing reports, construction planning documents, construction job records, and the summary construction information listing).

One of the key elements of ORR is to track changes in the execution of the project lifecycle. The change information is documented and tracked in the objects of design change requests (DCR) during engineering and construction change requests (CCR) during construction. An information data warehouse system has also the ability to document, track, and recall all the historical changes that have happened throughout the project lifecycle and stored in the warehouse. This capability is important to support efficient ORR activities.

ORR has direct inputs from SOT, which verifies that the plant process and facility systems meet the operational and safety requirements and that the plant is ready for start-up and operations (see Figure 8.4). Furthermore, ORR requires document evidences that there are sufficient qualified and trained management personnel and operators to run the plant safely. Operator training is a part of commissioning activities.

Figure 8.4 Operation Readiness Review (ORR)

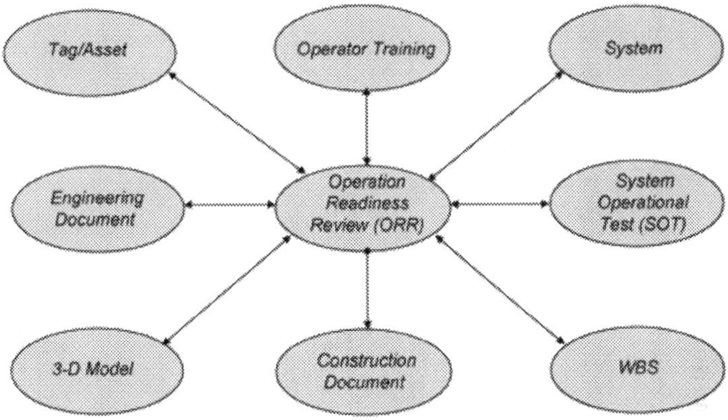

## 8.5 Mechanical Completion and Hand-over

After completion of SOT and ORR, the project phase of the process plant is done and the plant is ready for hand-over from the EPC to the owner or operator. There are many procedures and steps in the physical hand-over of the plant from subgroups to groups, from subsystems to systems and from inspection to final acceptance. However, from an information management point of view, information hand-over is just as important, if not more so, than the physical hand-over.

Hand-over is the dividing line or interface between the project phase and the operational phase. Because of this, there are traditionally sharp discontinuities and problems on information transfer. From a financial point of view, there are two different responsibilities from the EPC to the owner or operator. In addition to this, plant project capital expenditure is usually managed from a different plant operational budget. In a major process plant, experience shows that information clean-up and reset for successful hand-over can cost millions of dollars. Properly managed information data warehouses reduce this cost threat considerably.

Preparation for a successful information hand-over means having the right information in the right format at the right time, and it should start as early as possible in the engineering and design phase. The owner or operator should specify not only the plant products and operational and safety requirements but also the information requirements for operations and maintenance. These may include information types from specific documents, databases, 3-D models, and vendor data. It is also important to specify the information format so the information hand-over from the project phase can link with the various application programs for operations and maintenance.

In the previous chapters, use of a plant lifecycle information server has been promoted. This approach structures the various project lifecycle activities as a continuous process characterized by consistent information content and format. With this is in place, the hand-over is much more efficient because obtaining the correct information to satisfy the owner's or operator's requirements with a minimum of adjustment is much easier. Figure 8.5 shows how the hand-over object links to the basic engineering information objects, the procurement and vendor objects, the construction objects, and the commissioning objects. Figure 8.6 shows a summary of these objects, including the commissioning objects of systemization, system operational test and operational readiness review.

Figure 8.5 Turn Over

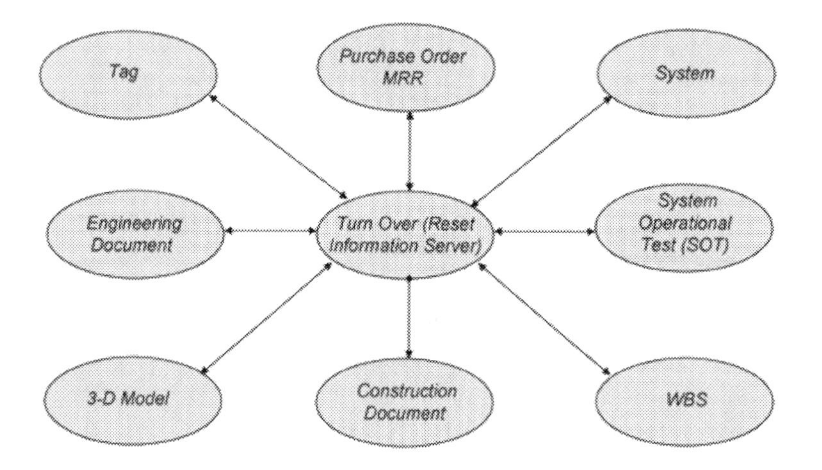


In the project phase, the information data warehouse contains a very large amount of information. The information objects are set to have their contents and linkages for efficient operation of different project steps. In the hand-over process, the objects are reset along with their contents and linkages in the information data warehouse to meet the different priorities and usage requirements of the plant operational phase. No information in the warehouse is, however, deleted that it can always be recalled for reference and future use.

# Figure 8.6 Mechanical Completion & Commissioning

Existing Objects

- Area
- System
- Discipline
- Tag
- Document
- 3-D Model
- WBS
- Purchase Order
- Vendor
- MRR
- Asset Number
- Construction Job Records
- Construction Acceptance Test
- Construction Information Listing

New Mechanical Completion & Commissioning Objects

- Systemization
- System Operational Test (SOT)
- ORR (Operation Readiness Review)
- Turn Over (Reset Information Server)

There are priorities for process-related information such as intelligent documents for P&IDs, control diagrams, instrument settings, electrical single lines, and so forth. There are databases for line summary, process equipment, instrument, and control summaries. There are linkages from tags to asset umbers and from equipment to 3-D models. For maintenance purposes, vendor data such as operational procedures, maintenance, and spare requirements are linked to equipment and component asset numbers. The results of SOT, which provides the basis for startup, are summarized for plant operations.

# CHAPTER 9
# PLANT FUNCTIONS

## 9.1 Overview

As discussed at the outset, plant functions in the process plant lifecycle are much more than the EPC or project functions in terms of schedule time and cost. A typical process plant can have a twenty to thirty year operational lifespan compared to the EPC phase of only two to three years. The yearly operational cost of a plant may be in the range of ten to fifteen percent of the total investment to design and build the plant, but the accumulated cost for the plant lifecycle operations can be five to six times higher. Therefore, it is very important to set up and manage the plant functions properly such that a small reduction of 5–10 percent yearly on operational costs can pay the total cost of building the plant.

The major plant functions are in operations and maintenance supported by functions in plant engineering, procurement, and cost control. There are plant compliances in QA/QC, configuration management, and satisfaction of regulatory safety and environmental requirements. Different types of process plants may have different detailed requirements for plant operations and maintenance but from an information management point of view there are essentially many of the same functions.

Figure 9.1 shows the basic information management plant functions. At the center is the plant information server that contains the plant as-built information generated by the EPC and organized during project hand-over for plant operations and maintenance. As operations goes forward, additional information on the plant is generated and added to the server. The server may also be connected to a separate document management system, which contains engineering and vendor documents and additional documents as generated by plant operations.

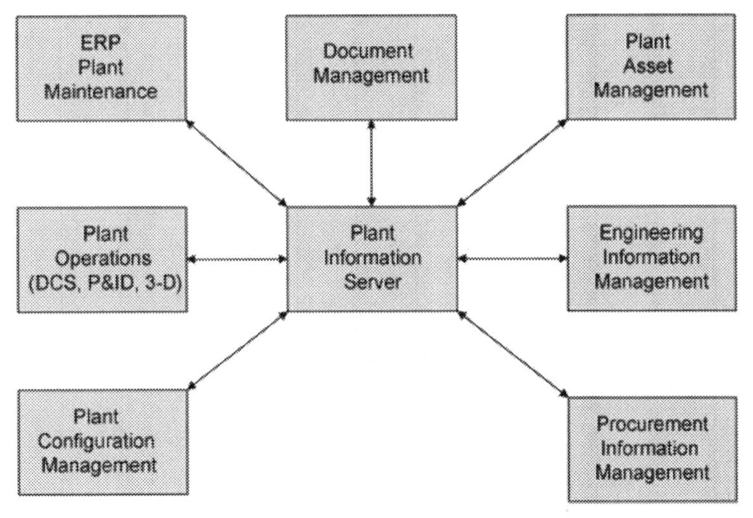


Plant operations also include engineering and procurement functions. Engineering manages and supports as-built engineering information on the server in the form of data, documents, and 3-D models and also generates new information as required for plant modifications. Procurement manages and supports vendor information and buys expendables, replacement spares, and new items.

Using a plant information server drastically improves the efficiency of plant operations. All engineering and operations-related information from process and engineering to asset items like equipment and components can be retrieved efficiently from the server. Availability of accurate information at the right time enhances the decision process in operations for trouble shooting as well as responses to operational safety and emergency conditions.

Plant asset management is one of the keys to good plant maintenance. Asset management means more than just physical assets and includes information management related to the assets. Asset items are linked to engineering through tag numbers and to databases that include vendor information on maintenance requirements, spare parts, warranty, and so forth.

Plant operations need to satisfy plant QA/QC, safety and environmental requirements, and plant configuration management. During operations, any plant changes—whether in engineering and design, operational procedures, or physical items—need to go through a documented request and approval process and then be stored in the information data warehouse server.

The basic goals of plant operations are to operate and maintain the plant and to produce the required process products efficiently and economically. But, plant operations also include searching for opportunities to improve both the quality and the quantity of the process products.

## 9.2 Plant Engineering and Procurement

Plant engineering begins with the support and management of engineering information in the information data warehouse server. As stated in the previous chapter, engineering information is divided into two major categories: process-related and facility-related. During the project hand-over, engineering works with operations to define operational information requirements. The major items related to process are mechanical processes (P&ID), instrument and control, electrical, process piping, and process equipment. Items related to facility are HVAC, mechanical equipment, electrical equipment, and mechanical utility. Civil, architectural, and structural information is also maintained in the information data warehouse server and can be retrieved easily for use.

There is integration or linkage of the three engineering information types: document, database, and 3-D model. Engineering documents can be stored directly in the information data warehouse server or, in some cases, in a separate document management system. Key words, such as document name, discipline, issue date, revision number, revision date, and so forth provide information linkages from the information server to the document system server.

Engineering maintains plant 3-D models and divides the composite models into submodels based on different regions of the plant and different process functions. Major components in the model are identified by tag numbers and are linked to databases and documents. Models can be used directly to track operational functions, to help troubleshooting and to track safety features. Models can also be structured and used for operator training. On maintenance, models can be linked with a maintenance program or programs for planning of maintenance steps and procedures.

Engineering maintains engineering information in the information server as a single source. When operations and maintenance connect and retrieve engineering information from the server, the information can be used with confidence that it has integrity which means with guaranteed accuracy and currency.

When plant changes or modifications are required, plant engineering because of its knowledge of the plant as-built conditions, conducts studies, prepares engineering change requirements and specifications, and provides support in engineering and design functions. For major revisions, plant engineering may act on behalf of the owner or operator to interface with an EPC contractor to provide as-built plant information and to review and approve new design changes.

Plant procurement provides procurement functions for plant expendables, spare part items, and any new equipment and items. There may be a separate procurement program, but the plant procurement requirements and functions are essentially the same as in the EPC phase with the issuance of requisitions, purchase orders, and acceptance of purchased items. Procurement interfaces with maintenance and the vendors on all vendor-supplied asset items to resolve any maintenance-related questions and issues.

On any new revised and purchased engineering items and components, procurement provides linkage from the engineering tag number to the assigned new asset number for operations and maintenance. All vendor-supplied information and documents on the new assets are tracked and stored in the information data warehouse server.

## 9.3 Operations

Once a process plant is built and in operation, the operational group is responsible for safe and efficient operation of the plant to produce the required process products. Plant operations are continuously monitored and recorded. The operational results are analyzed to look for potential process improvements. Process plants from many different process industries may have different operational requirements and procedures as each product requires, but there are still many similarities as discussed in the following.

In the control room of a typical process plant, various instruments and controls are designed, set, and used to monitor and control the process and facility functions. From an information management point of view, performance parameters of process functions are tracked and displayed in control consoles. Digital instruments and control settings are continuously monitored and recorded.

If the plant information is stored in an information data warehouse server, then there are computer terminals and monitors in the control room that are connected directly to the server. There are also terminals at selected locations throughout the plant at which users may connect to and communicate with the information data warehouse server. With this arrangement, information stored in the server can be retrieved and used efficiently by the control room operators and other users throughout the plant.

Under normal—and especially up-set or emergency conditions—operators can retrieve and look at any process-related information. For example, an operator can retrieve and view a P&ID and from line numbers retrieve any process and operations-related information. The operator can identify and retrieve information by clicking on tag numbers for equipment, instruments, or control valve items. These items can link to their datasheets and through linkage to asset numbers connect to their related vendor data such as operational procedures, settings, and so forth. Documents such as instrument loop diagrams, electrical single lines, and inter-connects can also be recalled to trace and solve operational problems.

Three-dimensional models are powerful tools for improving operational efficiency because any specific model can be retrieved and viewed in the control room. Tagged items in 3-D model such as valves, instruments, and equipment can be tracked to their corresponding P&IDs and datasheets. Process piping in the model can be identified through its line number and tag number and linked to the models and piping isos. For safety operations, closed circuit TV cameras can be installed at various locations in the physical plant. The views can then be linked to the corresponding 3-D models for interpretation and solution of operational and safety problems.

Facility-related operations such as motor control center (MCC), HVAC, fire protection, and so forth are also managed and viewed from the control room with facility-related information that can be retrieved efficiently from the plant information server. Facility 3-D models are also used to improve facility operational performance.

Operator training is a key function of operations. Three-dimensional models in connection with process and facility information from the plant information server can be used as very effective tools for training. Process systems, P&ID systems, control systems, electrical systems, and so forth can be explained, tracked, and illustrated through 3-D models. Operations have other traditional plant functions such as accounting, cost control, and information systems (IS) that manage network and hardware and software for computers and servers.

## 9.4 Maintenance

There are many good programs that are used for the maintenance of process plants that include functions such as maintenance planning, worker duties, worker assignments, asset management, records keeping, and so forth. From an information management point of view, the key issues are availability of accurate plant information for asset management and record keeping.

During hand-over, the plant maintenance staff would have worked with the EPC project staff to identify information requirements for maintenance. If an information data warehouse server is used, then such information is organized, assembled, and stored in the server. If this step is done properly, there are considerable cost savings because the maintenance staff does not have to reassemble the information to fit the content and format of the maintenance program.

The maintenance staff can retrieve information on an asset item such as date of installation, maintenance requirements, maintenance history, warranty period, availability of spares, and so forth. The staff communicates with vendors on any changes, maintenance problems, and any new items.

In a process plant with good information management, the maintenance staff can retrieve plant information conveniently from terminals that are installed at critical locations throughout the plant and are connected to the information data warehouse server. This approach provides efficiency and avoids costly mistakes from using incorrect or untimely information.

## 9.5 Plant Configuration Management

Plant operations and maintenance follow established requirements and procedures. Plant QA/QC monitors and controls that the plant is operated and maintained as specified. Plant configuration management assures that any plant changes go through the required request, approval, and implementation steps, including documentation of the change. Plant engineering, procurement, operations and maintenance collectively manage the plant configuration so that any changes can always be tracked accurately.

The key to configuration management is thorough record keeping. In operations, daily operational records in terms of instrument and control settings and process functional results are recorded and documented. Any abnormal or emergency conditions, along with any actions used to resolve such problems, are especially noted and recorded. This is important for identifying any potential faults in the design and operational procedures and also for tracking, analyzing, and preventing any repeat future occurrence.

All routine maintenance records are kept, and any movement and change of plant assets are especially recorded. Such information may also potentially help to analyze any operational and maintenance problems.

All operational and maintenance records may be kept in separate servers through an operations and maintenance program. Selected information may be linked to the information data warehouse server or can be stored directly on it. The historical operation and maintenance records should be kept for a determined period of time throughout the plant's lifecycle. Such records are important for safety and environmental compliance, for trouble shooting potential problems, and for future changes and modifications to the plant.

All engineering and procurement information related to the configuration change of the plant is stored in the information data warehouse server as part of the plant as-built information. Any change to operations and maintenance from those changes must also be recorded.

# CHAPTER 10
# ADVANTAGES OF PLANT LIFECYCLE INFORMATION MANAGEMENT

## 10.1 Why Plant Lifecycle Information Management

In the previous chapters, we have discussed the what, when, and how of information management for the various lifecycle stages of process plants. We have not mentioned in details 'Why' we have to do information management or what the advantages of plant lifecycle information management are. This chapter addresses that.

Because the basic advantages of information management are applicable to most of the plant lifecycle activities, we can summarize the advantages and present them here as a complete entity.

Historical data shows that throughout plant lifecycles there is considerable time and effort spent searching for and retrieving pertinent information. Modern technology provides tools that can help us to improve efficiency of these tasks. But, the key is still what information is required at what time and how we can organize the information to search and retrieve it effectively. This is where plant lifecycle information management becomes an especially powerful tool.

First, we know that critical information, once developed, is used over and over throughout the plant lifecycle. For example, one P&ID used for engineering of piping, instruments, and process equipment will also be used for procurement, construction, and operations and maintenance of the plant. Therefore, we want to organize that P&ID, its graphics, and its data as a single source of information to be used throughout the plant lifecycle without concern of informational redundancy.

Second, once we know where to get information efficiently, we want assurance of that information's integrity. Information integrity means that the information retrieved from the single source is accurate and current at the specified time. This is important because changes are inevitable, and we want to be able to manage changes and still maintain information integrity. We introduced earlier the concept of two levels of information: information in the working environment (WE) and information in the controlled environment (ICE). This approach reduces considerably any confusion of information integrity at a given time throughout the plant lifecycle. It allows any necessary work in any aspect of the plant functions to proceed without changing plant function until such work is published in ICE. Without a rigorous, two-tiered information management system, confusion and chaos would inevitably result as one discipline's efforts interfere with, or are not implemented, by another.

Third, we have mentioned the three major forms of information: documents, data and 3-D models. The implication is that the trend on information is more data-centric than document-centric. From an information management point of view, the data-centric allows much more flexibility, and much more efficient information linkage and integration that decreases the number of interfaces, increases coordination, integrates multiple applications, and increases the feasibility of automation. Automation is the foundation to higher efficiency and productivity.

Finally, information management provides the necessary tool for project and plant compliances on quality, safety, and regulatory requirements. With electronic or digital information and modern technology, we can track and store a very large quantity of information. Thus, we can afford to have long-term storage of historical information on both project and plant operations. By knowing what happened historically, we could identify problems, correct mistakes, and improve the overall project and plant efficiency.

## 10.2 Reduce Capital Expense (CAPEX) and Time to Market (TTM)

A successful project designs and builds a plant that produces the required products with minimum capital expense (CAPEX) and that reduces the time to market (TTM) to the shortest possible time. In other words, in a successful project one manages the schedule and costs most efficiently.

In engineering and design, the trend to reduce cost and increase productivity is to have more automation. The first step to this is more linkage and connection with the right information. For example, within the piping discipline, 3-D models of piping provide enough dimensional and components attribute information that one can not only construct a finite element model for piping stress analysis but also extract piping isometrics and materials for procurement, fabrication, and installation of the pipe. This inter-discipline linkage of information on automation from P&ID to instrumentation, process equipment and electrical provides data on instrumentation design, to generate process equipment datasheets, and to develop electrical single lines.

We can drastically reduce cost and shorten schedule if we can perform so-called 'once through' engineering; that is, to do engineering and design with essentially no changes. The basic approach to minimize changes is to use the correct information at the right time and to avoid the expensive correction of mistakes. (Also, the fewer the design changes introduced later, the greater the cost savings, since changes at the design point have the most far-reaching impacts on project implementation.)

An even more basic approach for minimizing changes involves spending enough resources at the pre-engineering or the front-end phase of process and engineering to look at many process alternatives and to study different engineering and design arrangements. Design changes would then be minimized because the right design would already have been examined in terms of trade-off results between several design alternatives.

Linkage of engineering information to procurement provides correct engineering information to automate and generate the necessary procurement documents. It is also important to obtain as early as possible the vendor data on purchased equipment and components such as size, dimensions, interface connections, and installation details. This minimizes the engineering design changes because of the late availability of wrong or different vendor information from what is specified in the design. This can be a difficult process, however, because of the bidding during the procurement process as required by either the owner or regulations on some projects.

With the modern technology, procurement can manage a very large, worldwide vendor information database. This improves on the selection of qualified vendors for bidding and procurement. Procurement has more advanced information on the availability and pricing trend on commodity items, equipment, and component items; this information can be fed to engineering to make the most strategic selection for the design.

Easy access of single source engineering design information improves construction planning and the actual construction process. The early involvement of construction not only improves engineering and design of the plant but also improves the construction cost and schedule. Information on constructability helps engineering to select the best design alternatives. Early studies on construction sequences, construction equipment requirements, and construction yard layout all reduce construction costs therefore the capital expense of a plant.

From the above, the cost advantages of using single-source, controlled project and plant information stored in an information data warehouse server that can be used over and over throughout the project lifecycle activities should be clear. Such an approach ensures information integrity, avoids information redundancy, and minimizes mistakes.

## 10.3 Reduce Operation Expense (OPEX) and Maximize Time in Market (TIM)

As we have mentioned earlier, a small percentage reduction of annual operational expense (OPEX) can accumulate into a substantial sum over time. Other than scheduled plant shut downs for maintenance, a plant should operate continuously at full production capacity. In other words, good plant operations and maintenance should ensure maximum up time and thus a maximum supply of products to the market (TIM).

The proper hand-over of the right information sets the stage for efficient operations and maintenance. For operations, we want required engineering information to be organized as a single source so it can be retrieved efficiently in the control room and at critical locations throughout the plant. This is particularly important for all safety and hazard-related information to ensure a minimum of operational interruption in abnormal or emergency conditions.

As the plant moves into normal operation, the operators should continuously look for improvement of operational efficiency. Operational data are collected and analyzed so that engineering information may be rearranged, additional linkages may be added, extra terminals can be installed, or new operational procedures may be implemented.

Vendor data on equipment, instruments, and components as asset information databases are organized to provide inputs for maintenance planning and activities. Normal replacement of plant equipment and components are included as part of scheduled maintenance to improve plant reliability and TIM. But there is a cost trade-off between reparability versus replacement of components. Asset information management gives the information inputs on equipment warranty, designed life expectancy, and so forth to help make the right cost trade-off decision in these circumstances.

## 10.4 Optimize Operating Parameters (OOP) to Increase Productivity and Outputs

In addition to increasing productivity in normal operations and maintenance activities, there is the basic requirement to improve quality and quantity of the products. Collected operational data gives the necessary information to optimize operating parameters (OOP). Operating parameters might be adjusted or the ratio of the input process components could be rearranged. Performance of the process equipment is studied to evaluate whether the equipment is operating at the designed level or if equipment operating parameters could be adjusted to improve efficiency. The overall goal is to reduce operating costs or to increase the overall product quantity and quality.

Over time, the plant capacity may need expansion or the plant process may change because of new process development information. In either case, the starting point will be to conduct process studies and then, from the new or revised process, re-engineer and re-design the plant accordingly. This will require having good and accurate information on the plant's as-built conditions—information that is available in the information data warehouse.

To ensure accurate as-built plant information requires a continuous process of record keeping to track daily operations and maintenance from the first day of hand-over. As such, process-related information is important not only for operation of the plant but also for plant changes because accurate information on every detail of the physical components of the plant will be required for re-design. It is the field engineering's responsibility to keep up with the plant changes and to maintain the as-built conditions. If there is a plant information server, the documents, databases, and 3-D models are all maintained in the current and correct positions. With information management, historical plant information on plant changes is maintained and can be retrieved and used as inputs for major plant revisions and modernization.

OOP is an important factor in increasing productivity and the outputs of a plant. It expands the usefulness and extends the life of a plant. Information management through an information data warehouse is the key to a successful optimization and expansion of plant operating parameters.

## 10.5 Satisfy Project and Plant Configuration Management – Safety and Compliance

Information management provides necessary tools to satisfy project and plant configuration management for quality, safety, changes, and compliance. To manage project and plant configuration efficiently can produce major cost savings. On the other hand, failure to satisfy project and plant compliance can cause very high expenses and schedule delays.

Throughout project lifecycle, we can identify, track, and link safety related information specifically in documents, data, and 3-D models. For example, safety control features in a P&ID can be shown in a 3-D model and linked to a safety procedure document. Using this approach as part of information management improves the engineering design process for safety and enhances quick responses to safety-related issues during operation.

As stated earlier, to show compliance with safety and environmental requirements, we can generate databases that link engineering design information with design criteria and regulatory documents. For example, to obtain operating license for a process plant like a pharmaceutical plant or a nuclear power plant requires extensive documentation that demonstrates evidence of design and construction compliance. Information management provides a logical approach, convenience, and high productivity in performance of these tasks.

Change management is always a major function in either project or plant lifecycles. The ability to record, track, and recall change information efficiently through organized databases is a valuable tool for configuration management. Long-term change information storage and retrieval is another plus for cost control in change management.

# CHAPTER 11 IMPLEMENTATION OF PLANT LIFECYCLE INFORMATION MANAGEMENT

## 11.1 Overview

In the previous chapters, we have explained the principles and approaches for the what, when, and how of various stages of process plant lifecycle information management. In the last chapter, we have presented the many benefits of such applications. As a summary of this book, we describe in this chapter the key elements that are necessary for successful implementations of plant lifecycle information management.

Information management has long been established as a powerful tool, but it has only been in the last few years that it has been applied through the use of information data warehouses to the lifecycle of process plants. Owners and operators were the first to implement the technology to support their existing operations and maintenance programs. With the development of software from major technology application companies, more and more EPC companies came to implement the tools as well, at first for various parts of engineering designs and then later for procurement and construction.

Information management becomes almost necessary for integration and coordination of information for project and construction management on large projects where there are many engineering and construction companies. In some cases, owners or operators specified and directed the use of information management on projects because EPC companies were reluctant or slow to implement them due to cost considerations.

However, the trend is for more and more implementations of lifecycle information management in process plants because the benefits are becoming more apparent not just for the individual company in the lifecycle but for the whole project and plant.

Effective information management requires technology in computers, servers, network, database and application software. To implement the relatively new technology of plant lifecycle information management, the general first approach is to obtain information from the technology company that develops the information management software. A demonstration of the products and conducting a pilot project follows, if necessary. There are some cost trade-off studies to show the advantages of using the technology before a decision is made on implementing it.

In addition to technology, two other major considerations must also be addressed for successful implementation: a people and organizational element, and a work process element. Because of enthusiasm for the promise of the new technology, there is a tendency to put too much emphasis on it, so that sometimes we fail to understand the importance of these other two elements.

## 11.2 Technology

The database is the basic technology tool used to manage information. In engineering, especially in the process-related disciplines, tags with attributes in databases are traditionally organized to represent process equipment, valves, instruments, and so forth. In addition, process lines in P&IDs, electrical lines, and instrument loops, can also be labeled as tags. In the physical disciplines, such as piping and structural, we can also use tags to identify piping isometrics, structural supports, facility equipment, and so forth.

After procurement and during construction, most tagged items are changed from tag numbers to asset numbers. The tag database for an item is expanded to include additional information that includes an asset's purchase information and physical properties. For example, a process pump asset database may include purchase order number, manufacturer's name, model number, serial number, the installed location in the plant, and so forth. Asset numbers are used, for example in maintenance for service requirements and in operations for trouble shooting. Tag numbers are linked to their appropriate asset number through databases and while an asset number is unique, there can be multiple asset numbers assigned to one tag number.

An information data warehouse is a relatively new technology tool for information management. It manages information integrity (accuracy and currency) from a single source throughout the plant lifecycle. It provides tools for information efficiency on storage, distribution, viewing, and retrieving. It organizes project and plant relationships for effective tracking of information.

The most important function of the data warehouse is integration or linkage of database information at one location. Given the three basic information elements—of document, database, and 3-D model—we can use the information data warehouse directly as a document management system, or we can use database and link the data warehouse to an external document management system. The same database approach can be used to link 3-D models with the information data warehouse.

Tag number is the key used to identify and link the same information element from database, document, and 3-D model. Successful information data warehouse software may be judged by how effectively and reliably it can mark or identify the tag number in the three information elements.

As mentioned earlier, (paper) documents are still used extensively in the process plant business. Many documents can be produced reliably with information integrity from data information stored in the data warehouse. Such technology uses an intelligent template approach. For example, from P&ID database information, line summary and process equipment summary documents can be produced. From valve and instrument tags, datasheets and specifications can be produced. In procurement, templates can be developed to produce requisition and purchase order documents through engineering databases in the data warehouse.

Throughout the plant lifecycle, many of the existing application programs are used. In engineering, there are P&ID and instrument design and piping and structural 3-D design programs. Many good programs are available for use in procurement. In construction, there are programs for construction planning, progress reporting, construction verification, and testing. Depending on different process industries, there are established operations and maintenance programs. Information management effectively links these application programs to an information data warehouse for overall plant lifecycle applications.

There has been some success linking specific outputs from one application program directly as inputs into another program. But, it presents two major problems as well. First, such application requires much more complex programming in addition to being less reliable and difficult to apply. By analogy, just as something gets lost when translating from one language to another, the same risk is here, except that for plant process design, only a no-error translation is acceptable. Second, database linkage is less flexible for changes that can be applied to different project requirements. This may result in implementation costs that are too high for general applications to various process industrials.

By contrast, a single-source information data warehouse is generally a cost-effective approach. Output data from an application program uploaded to the information data warehouse can be used as inputs to several application programs. Information does not need to be transferred in and out of the data warehouse automatically. A simple spreadsheet type of database program can be used manually to upload the data. For example, line summary information generated in a P&ID program can be assembled and setup as a spreadsheet and downloaded to the information data warehouse as attributes from a group of line tags. Instrumentation and piping can retrieve and upload the required attributes from the line summary spreadsheet as inputs to their design application programs. The approach may seem cumbersome, but experience has shown that the approach is simple, easy to implement, and flexible to fit different applications and reacts easily to changes.

## 11.3 People and Organization

The human element in a successful information data warehouse implementation applies to all people from the top down. From top management, vision and commitment is essential. Vision means understanding new business trends, the owner's or operator's requirements, and the potential benefits of lifecycle information management. Commitment means not just setting goals and financial guidelines for the project but providing support to see through the expected ups and downs of implementing a new technology. For the people in project management, they must understand that the benefits of information management are for the whole project and not just for one or two functional groups. Project management may have to adjust the traditional budget allocations such that there may be more setup costs and higher than usual efforts in the upstream disciplines so that the downstream disciplines benefit. The goal is to provide right information at project hand-over and to save costs for owners or operators on operations and maintenance.

For people in functional discipline groups, communication to understand implementation requirements and training are crucial. Training is always a difficult process because two to three weeks of training is not sufficient to make a worker proficient on using a new technology. Since users have different backgrounds and technical abilities, one approach is to identify super users and give them incentives and additional training so they become leaders of the new user's group. Super users need to demonstrate not only technical ability but also an understanding and willingness to learn and use the technology. Another approach is to hire new users who already know the new technology as core user members. In a tight labor market, one may have to pay premium for this kind of help.

It is human nature to resist change when it is implemented without taking into account the person being asked to change. In addition to technical training, then, some programs are required to condition worker's attitude and to build their confidence on how to use the new technology. New procedures are also required for the users to follow on new work processes that are required for the technology. This is all in step with management's commitment and support for successful implementation on a project.

Since it is a new implementation, sufficient IT staff needs also to be trained and to learn how to maintain and support the technology. In addition, there are organizational issues that need to be addressed. Management should emphasize the importance of a team effort on various organizations and define clearly the areas of responsibilities that may not be the same as the traditional ways. Experience shows that it may be necessary to assign a group or to organize a committee, which has the authority to resolve conflicts among the participating groups.

Many owner or operator companies traditionally have two separate organizations: one for the capital expenditure for design and building of the process plant and another for operations and maintenance of the plant. In this case, there are potential conflicts because different budgets may impact different performance requirements. The preferred approach is that the two organizations should have close coordination to take advantage of the lifecycle information management. In that sense, plant-operating organizations should participate as early as possible with the engineering and construction company or companies. As a matter of fact, the owner or operator can specify that part of the deliverables at the end of the project will include a project information data warehouse that is organized to contain all the necessary and correct information for operations and maintenance of the plant. This drastically reduces the cost of project hand-over.

## 11.4 Work Processes

Throughout the years, process plant owners and operators, engineering and design firms, and construction companies have developed methods and procedures on how to perform their work. Organizations from management down to project and functional units rely heavily on traditional approaches and the experiences of each individual company. This is the reason that the most difficult part of implementing a new technology is the necessary requirement to change the work process.

The usual approach is to apply the new technology directly on the traditional work processes. This frequently fails because we don't know what work process changes are required to make the new technology work.

Furthermore, work process changes require a trial and error method before the best application for the technology can be found. Then, what can we do? We can get help from the company that supplies the technology software. We can gain knowledge and experience from other companies that have successfully implemented the technology. We can hire technical consultants who are familiar with the technology of information management and required work processes. The main point is that we should be open-minded and follow the slogan that we cannot force the technology to fit our traditional work process but must change the work process to take advantage of the new technology.

Intra-discipline integration of information is more straightforward because communication within the discipline group can be managed more easily. On the other hand, inter-discipline information management requires a more detailed definition of input and output information relationships from each step of plant lifecycle. For example, the mechanical process group produces P&ID output information—whether it is in document or database format—to satisfy the input requirements of the instrument, piping and process equipment disciplines. The P&ID outputs should also satisfy the procurement and operational requirements for the plant owner or operator.

For extensive use of databases, attributes for each database should also be clearly defined. For example, we know what the attributes are for an instrument tag, a valve tag, or an equipment tag. We define also the attribute information for linkages of databases (e.g., a purchase requisition document's linkage to several engineering databases or a design application program to a 3-D model). These are all part of the 'what' for information requirements in the work process for lifecycle information management.

New standards and procedures need to be developed for the new changed work processes. Examples of this include learning when and how to download and upload information from the information data warehouse, knowing what attribute information to put in a database, knowing how to extract a document from an information template, and learning how to satisfy information integrity for information compliance. These are all part of work process changes.

## 11.5 Some Remarks

Computer-assisted design (CAD) was first developed to increase efficiency for producing drawings. It was not foreseen then how CAD would create powerful new advantages for engineering design. Similarly, word processing programs were developed to speed up typing, and its changes to the office work process were not foreseen; in one sense, instead of making the typist's job easier, it eliminated the need for a dedicated typist entirely. The Internet also has so revolutionized how we communicate that the world is no longer the same.

Plant lifecycle information management is a relatively new technical tool that has the potential to increase both project and plant operations to much higher levels of productivity. As we proceed with more and more implementations, we may discover many new areas of applications that may change how we design, build, and operate process plants.

# APPENDIX A
# OTHER TOPICS

## A1 Knowledge Management

Knowledge may be thought of as an accumulation of facts and principles on a certain subject gathered by learning, actions, and experience such that expertise is developed on the subject. In the many years of successful execution in design, construction, and operations of process plants, knowledge has been gained by the process industry on the cost-effective and expeditious way to design, build and operate plants. As new processes and technology develops and refines, new knowledge expands continuously. Information gathering and management is the first step for knowledge development. In general, some basic knowledge in the process industry is gathered and represented by the following topics: industry standards, best practices, work procedures, and experiences.

The process industry summarizes and publishes commonly held knowledge as standards for use by the industry. These standards—developed by owners or operators, EPC companies, technical societies, and universities—are important because they provide guidelines for the execution of projects and the operation of plants. This is a cost-effective approach for applying commonly held knowledge without each organization developing on its own. There are standards on engineering designs, environmental and safety requirements, and acceptable plant operations. Some developed standards become industrial and governmental regulations, or even licensing requirements.

Industrial best practices are developed as further expansion of knowledge on standards. For different process industries, there are best practices for engineering, procurement, construction, and operations. Based on knowledge and experience, best practices include instructions on how to perform specific tasks. For example, in engineering, there may be application of a given technology or software for design or guidelines to set up a recommended work process. Again, best practices are cost-saving approaches because they are efficient applications of knowledge.

In addition to industrial standards, EPC companies establish company-specific procedures as guidelines for performing certain project and plant lifecycle tasks. These procedures are important for several reasons. By following them, workers can better coordinate through common approaches to perform tasks with consistent results. Owners or operators can more easily review procedures of EPC companies to gain some assurance that the company can perform the tasks to follow the industrial standards with expected results in quality.

What make an EPC company good is its many years of experience in engineering, procurement, and construction tasks. What makes an owner or operator good in maintaining and operating a plant is also its many years of experience in doing such tasks. The important point is how to manage such knowledge through information and experience so that standards, best practices, and procedures are consistently produced. Doing these result in an accumulation of knowledge that, when applied, provides a more cost effective approach to designing, constructing, and operating a process plant.

## A2 Plant Design Template

From knowledge and experience, plant subsystem and system design templates can be developed for a specific process plant. The templates can then be applied with efficiency on similar or identical plants. The templates can be process or P&ID diagrams or part of physical plant for piping layout and standard equipment support. To take this one step further, a design template for a process train or even a complete process plant can be developed.

The plant template approach has been applied successfully in pharmaceutical and food process industry. Once a standard process plant is designed, it can be site adapted domestically or at foreign locations. In the power industry, since there are only few major equipment suppliers of generators and turbines, it is popular to have a standard design for cogeneration power plants for different power capacities.

For the major nuclear power reactor suppliers, the trend has been to obtain a license for a standard design that can be site adapted even for the most severe site environmental conditions like hurricanes and earthquakes. While it may cost more to strengthen the plant structures and equipment and piping supports this approach avoids the very lengthy process of obtaining a license for each specific site condition.

For an oil refinery or a wastewater treatment plant, the plant typically occupies a large area. It is not always possible to find the site to fit the layout of a standard plant design. If the plant equipment layout and major piping arrangement need to be changed, then we essentially lose the advantages of a standard plant design template. Another problem is that if there are many choices of major equipment and components, site adaptation of a process plant can also be difficult because differences in purchased items may require different plant layouts.

However, there are many schedule and cost benefits in site adapting a standard process plant design. At minimum, best practices on standard designs for processes, P&IDs, and plant layout of standard subsystems and systems can be developed so they can be selected and arranged for different process plant applications.

## A3 Workshare and Virtual Office

In the EPC business, there has always been normal information exchange between offices, the field, and clients. With rapid advancement in computer, electronic, and digital technologies, information transfer has changed mostly from paper-based documents sent through the mail and by fax to more extensively online review, comment, and revision of electronic documents, especially in the engineering and technical field.

Beginning in the early-1990s, the global competition of process business put cost pressure on US-based EPC companies to revise their traditionally set-up of a centralized engineering office. This led to the development of workshare approaches as a low-cost solution. Engineering centers were created to perform engineering and design work in the other countries such as India and the Philippines.

From the owner or operator point of view, especially on government projects, the trend in the last ten years has been to demand more and more execution of major engineering and design activities at or near the project sites. This puts a major strain on the traditional EPC engineering organization. The ability to assign good people to the remote job site, to hire qualified persons locally, to distribute and manage work procedures and processes, and to pay premium uplift salaries are all problems. Workshare seemed to be a good solution.

Through computer and network technology, the workshare has been developed successfully by using the technical approach of 'replication'. As shown in Figure A1, there are two work sites: location A and location B. Each site location has its own workstations, servers, and application software connected through a local area network (LAN). The working server contains data in the working environment (WE) while information that needs to be shared is published to the information server in the information controlled environment (ICE). Note that WE and ICE can be partitioned on the same physical server. The basic replication methodology involves duplication and synchronization of information server contents between location A and location B through the wide area network (WAN). If the workers tried to use the same design information from both locations at the same time, traffic on the WAN would be high and a very wideband network would be needed. An information replication workshare application alleviates most demand on network bandwidth, because replication is done during off hours with the potential of no more than eight hours out of synchronization between the two sites. If necessary, there can still be limited communications between the working servers to resolve any problems.

Figure A.1 Network & Server Setup - Replication

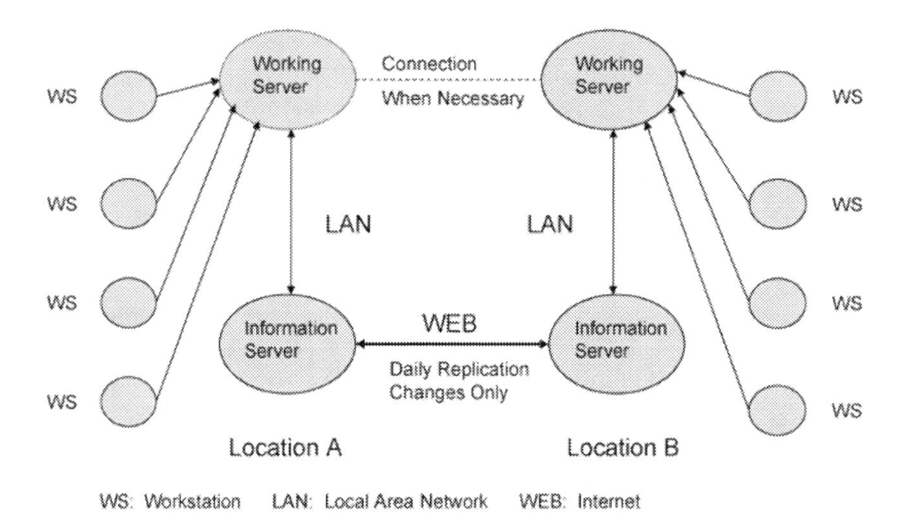


The workshare through replication can be expended to cover a large project with several project locations as shown in Figure A2. Information is replicated from the server at each site to a server at a main location. As there may be no replication between the servers from each site, information all flows through the main server. The main server contains information of the complete project and each local site extracts information of the other sites through the main server.

## Figure A.2 Network & Server Setup – Replication Multiple Locations

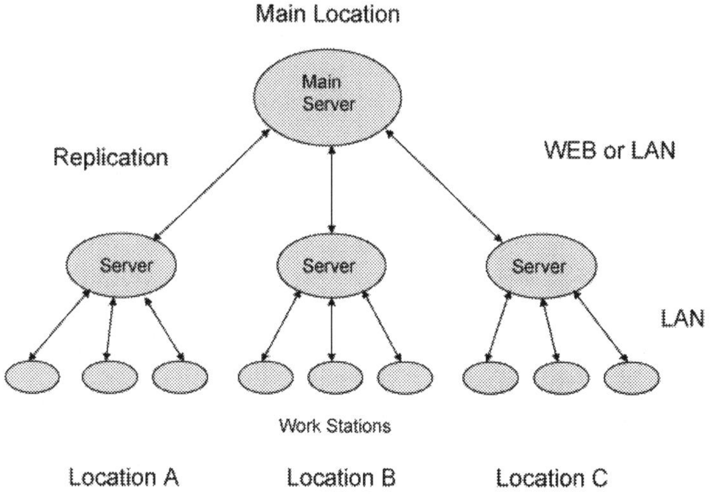

Current workshare technologies employ a 'virtualization' approach. Special software is installed on the workstations in each site, so the application program can communicate directly through network with servers at other sites involved with the workshare. The complete system performs virtually as if all components were in one location. Virtualization requires multiple servers connected through a LAN at one centralized location. This provides powerful storage and search capabilities and efficient management of large quantity of information. While replication is not required, one does need fast and large servers and an efficient broadband network.

Workshare requires considerations of client and project requirements, project scope, and available technology. The key point is that workshare involves management of information. To make a successful workshare project, we have to understand and define what and when information needs to be shared at the various project locations.

Workshare has become a standard business model in the process EPC industry for executing a concurrent design project. Workshare occurs among national and international design centers within an EPC as well as between cooperating EPC business entities. To facilitate design reviews and information exchange, more advanced workshare scenarios go beyond engineering and extend access of appropriate project data to clients, vendors, and subcontractors.

Using some of the workshare technology, EPC companies currently execute their various work processes that permit isolation of design subsets and then transfer responsibility for the subset from one group to another group. Worksharing has proven integral for EPCs because it gives them ability to successfully compete by leveraging specialized capabilities among projects, efficiently balancing work loads among their design centers, and providing efficient and effective interface tools for clients, vendors, and subcontractors without regard to restrictions on their physical location.

## A4 Design and Engineering Driven by Procurement

In a typical process plant, 30–40 percent of the project cost is in purchased items from tagged process equipment, instrument and control components, support equipment, and commodity items. Any cost savings in the procurement to pay suppliers are direct savings to the overall project. However, if we can minimize changes in the procurement process—and changes in the plant design process due to configuration changes of procured items—then these represent even more cost savings to the total project.

Process equipment specifications are determined from process design and P&IDs. Detailed design then proceeds on the basis of estimated process equipment configurations, which are derived from previous experience and data from similar plants. From the basic process equipment configurations, the other plant configuration for process piping and structural support are designed. The design of support equipment configurations and buildings then follows.

Unfortunately, the estimated process equipment configurations and dimensions of some of the tagged items and support equipment are usually not the same as the actual purchased items. This can cause major redesign of the plant with additional schedule delays and costs. Another problem is that some procurement process requires competitive bidding to select the vendor, which causes further delay before the actual configuration of the purchased equipment and items can be determined.

One approach to solve the problem is to specify the interface dimensions as part of the procurement specifications. This may not always work because it limits the choices on available standard items or may increase the cost of procurement because such units might be considered specially furnished items. Again, considerable benefits can be obtained if extra effort is spent in the process and preliminary design stage to examine design options and to fix the major equipment configurations so there are minimal subsequent changes in the final design. We can determine the process equipment specifications with more details and start the procurement as early as possible at this stage.

A trend in EPC companies is to have pre-qualified suppliers and even in some cases to have partnership agreements with the suppliers. In this approach, procurement has advanced configuration and pricing information on many commonly used items so engineering can design and pre-select the item with confidence that it fits the required specification. Caution must be exercised here, however, as potential conflicts of interest can result.

With a worldwide ability to purchase commodities and common support equipment and items, procurement can maintain a large database and keep it current on the availability and price of these items. This information is fed to engineering so the plant can be designed to fit these available items. The results are a more cost-efficiently designed plant with potentially good availability for any purchased items.

# APPENDIX B
# BUILDING/FACILITY LIFECYCLE INFORMATION MANAGEMENT

## B1 Building and Facility Types

In terms of numbers and dollar values, there are many more buildings and facilities than process plants. The building and facility lifecycles follow essentially the same pattern as the process plant lifecycle from design, procurement, construction, to operations and maintenance. The building or facility does not usually produce a product but it does, however, perform certain functions. Following are some examples of building or facility types grouped together based on their functions.

- Educational: these include facilities from grade school, high school, to major colleges and universities: Schools are designed with classrooms, auditoriums, cafeterias, sports and research facilities to serve an overall educational function.
- Hotels: in addition to providing lodging and food services for travelers, these also serve as casinos, conference sites, or resorts. Such facilities will have additional special setups and equipment to serve those functions.
- Transportation Facilities: these include train or subway stations, airport terminals, and cruise ship docks, and so forth. Special designs are required to accommodate these kinds of facilities.
- Commercial Real Estate: these include office buildings, large real estate developments, condominium buildings, and residential home complexes. These buildings and facilities support cities and communities.

- Government: these include post offices, state courts, city halls, army barracks, air force bases, and many federal, state, and military buildings. These buildings and facilities serve specific functions to support the nation and its citizens.

There are many other types of buildings and facilities. The point is that we have far more buildings and facilities than process plants. If we can follow the same lifecycle information approach as applied to process plants to manage information on buildings and facilities, then we have the potential to increase productivity and to reduce capital costs and operational expenses. Considering the size and number of buildings and facilities, the potential lifecycle cost savings from information management could well be in the billions of dollars.

## B2 Architecture – the Lead Discipline

In the building and facility industry, the abbreviation AEC (Architecture, Engineering, and Construction) is used to represent companies that run these project functions. As mentioned earlier, the common engineering feature that defines process plants is the P&ID, which translates the process requirements into engineering terms for design of the plant. In the AEC business, architecture is the lead discipline that communicates with the owner and translates the functional requirements into layouts, which through engineering and construction become buildings and facilities. For example, in an airport terminal, the traffic flow of passengers and an area for check-in through airline counters are some of the major design considerations. In a casino or resort hotel, the main floor is designed to accommodate the functional requirements of gambling equipment and actions. The engineering disciplines of civil, structural, electrical, and mechanical then work on the architectural layouts to transform these functional requirements into designs.

Currently, compared to the process plant design the AEC industry is much slower to convert from document-based information to data-based information. With the advancement of CAD, there are more intelligent architectural CAD drawings that can provide materials data and equipment summaries and also transfer and link such information electronically with other disciplines. Traditionally, a stand-alone company with its specific reputation and prestige has provided architectural services. There are architectural firms that are famous for designing hospitals, museums, high-rise buildings, and so forth, and they usually require support and coordination from separate structural, electrical, and mechanical engineering design firms. In this case, information transfer and integration are more difficult from company to company and the potential advantages of project lifecycle information flow cannot be fully realized. Construction companies also require further information transfer and integration with architectural and engineering companies. In a large project, the owner may hire a major project or construction management company to manage and integrate the project lifecycle functions of AEC companies.

## B3 3-D Design of Buildings and Facilities

As more software packages for 3-D design of buildings and facilities become available, the benefits of 3-D modeling for process plants is ever more obvious now for AEC ventures. Three-dimensional models provide better external visualization of the building configuration and can be used more effectively for presentation and review with the owner. Better selection and trade-off of internal design can also be achieved through 3-D color presentations.

Intelligent 3-D models contain attribute database information on the building components that can be transferred directly to 3-D structural design and analysis. In the mechanical engineering, 3-D room volumes give more accurate information for design of energy, heating, air conditioning, and air flow requirements. Room illumination and lighting requirements can also be better designed with 3-D model for the electrical discipline.

Following the process plant approach, tags can be used more extensively to identify building components (doors, windows, and so forth.), structural elements (beams, columns, and so forth.), functional equipment (desks and chairs in an office building, and so forth.) and facility equipment (elevators, firewater pumps, and so forth.). These tags can be linked to their datasheets and specifications and can then be fed directly to procurement. Three-dimensional models can be used for construction planning, construction review, and construction status reporting. In addition to the traditional documents, and databases, the third information type of 3-D models is available now to AEC businesses.

Owners have always had the facility management software tools to do building and facility operations and management. There-dimensional models add another tool to do these tasks more efficiently. We now have all the necessary tools to do full lifecycle information management of buildings and facilities through an information data warehouse application.

## B4 Building Information Modeling (BIM)

In the last several years, building information modeling (BIM) has been promoted to define the use of 3-D application programs for the design of buildings and facilities. As it is now almost fully accepted by the AEC industry, it is more and more specified as a requirement by some owners and government agencies. The use of the term BIM has even been expended to cover 3-D building site layouts for large projects and city planning.

BIM recognizes the information advantages of 3-D modeling and the linkage of data from architectural design to structural, mechanical, and other disciplines. There are, however, some issues with using BIM in the AEC industry. The inherent separation of the architectural company from other engineering companies presents legal responsibilities when transferring information from 3-D models. There are problems regarding how the many smaller engineering companies can adapt and fit into the BIM approach. Beyond this, there remain other restrictions to using 3-D models for procurement, construction, and even for owner's operations and maintenance functions, but those are topics for another book.

We would propose that BIM be revised to mean building information management, rather than building information modeling. The AEC business is the same as the process plant industry in that there are still the three basic information types (documents, databases, and 3-D models). Information management deals with integration, coordination, and transferring of these types of information. BIM can be expanded to include not only 3-D modeling but also overall lifecycle information management of buildings and facilities from architectural and engineering design, to procurement, construction, and to owner's functions of operations and maintenance. Documents can be extracted from 3-D models and transferred from discipline to discipline. Databases can be organized from documents and 3-D models for procurement and construction applications. Owners would then have the flexibility to use any type of information to achieve the most efficient approach for operations and maintenance. From this, it is clear that there are indeed many cost advantages of using BIM and building or facility lifecycle information management.

Robert Yang has over thirty-five years of experience in engineering, CAD and engineering systems, engineering management, and automation and CIE applications. He has worked as a mechanical and structural engineering expert on projects for petro-chemical plants, power plants, and infrastructures, specialized government, and military facilities.

Over the last seven years, he has been involved in the study, development, and implementation of information management systems for plant and facility lifecycles, and in particular implementation of information data warehouses. In this regard, he has worked as a project manager and technical consultant for the engineering, design, procurement, construction, and operations and maintenance for a major chemical process plant and a nuclear waste processing plant.

He has BS and MS degrees from the University of California at Los Angeles and is a registered mechanical and civil engineer in the State of California.

www.RCY111@aol.com